**Electric Traction
on the Pennsylvania Railroad
1895–1968**

Published with the cooperation and support of
the Association of American Railroads and
the Pennsylvania Historical and Museum Commission

Electric Traction on the Pennsylvania Railroad, 1895-1968

Michael Bezilla

The Pennsylvania State University Press
University Park and London

Library of Congress Cataloging in Publication Data

Bezilla, Michael.
  Electric traction on the Pennsylvania Railroad,
1895–1968.

  Includes bibliographical references and index.
  1. Railroads–United States–Electrification.
2. Pennsylvania Railroad.   I.   Title.
TF859.Z5P462   1979   385'.312   79-65858
ISBN 0-271-00241-7

Acknowledgments

••

Anyone who attempts to write about railroad history faces a dilemma: how to please the historian without alienating the rail buff, or vice versa. In this book, I have made an effort to appeal to both interests, realizing that I risk satisfying neither. The scholar, for example, may wish for less operating detail and a more analytic assessment of electrification's results. The enthusiast, on the other hand, may desire fewer discussions of corporate strategy and more emphasis on technical detail. My response to the enthusiasts is that many railfan publications already present the kind of information they are seeking. Rail hobbyists must now begin to broaden their horizons and acquire more than a superficial understanding of how railroads utilized new technology. To professional historians, I can only say that this is a selective history. Its aim is to synthesize the diverse technological, economic, political, social, and, yes, romantic elements of railway electrification into a coherent, readable narrative. It does not purport to be an all-encompassing history of any one of these elements. A prodigious amount of work awaits those who would chronicle the history of electric traction and twentieth-century railroad technology in general. This book represents no more than a first step in that regard.

To Professors Hugo A. Meier and Gerald G. Eggert of the Department of History of The Pennsylvania State University, where this book began as a doctoral thesis, I extend my appreciation for their careful reading of the manuscript as well as their many helpful suggestions for its improvement. The following individuals at Penn State also read the manuscript and offered valuable criticism: Philip S. Klein, professor emeritus of American history; W. E. Meyer, professor emeritus of mechanical engineering; and John C. Spychalski, professor of business logistics. Richard W. Barsness, dean of Lehigh University's College of Business and Economics, also read the manuscript and provided useful advice.

Recognition is also due the Association of American Railroads, whose financial grant enabled me to advance the work more swiftly and comprehensively than otherwise would have been possible, and the Pennsylvania Transportation Institute, which supplied me with a variety of financial and material assistance. I am especially grateful to the Institute's director, Thomas D. Larson, and its Transportation Systems program director, Joseph L. Carroll, for their encouragement of this project.

The cooperation of several other individuals and organizations has contributed to my work. Specifically, I wish to thank the Consolidated Rail Corporation, especially E. T. Harley, W. E. Kelley, and T. J. Judge, as well as H. J. Meehan and his staff at the Merion Avenue Records Center; the General Electric Company's Transportation Systems Business Division, especially A. P. Engel and B. F. Anthony; John MacLeod, librarian of the Association of American Railroad's Economics and Finance Department; and Bert Pennypacker and William D. Middleton, both accomplished writers in the field of railroad and electric traction history. I am also grateful to the staffs of the Engineering Library of The Pennsylvania State University, the University's Inter-Library Loan Department, the Industrial and Social Branch of the National Archives (Washington), and the Altoona Public Library.

Finally, my wife Deborah not only typed several versions of the manuscript, but on occasion served valiantly if not enthusiastically as research assistant. Only she can know the debt I owe to her.

Contents

Introduction

••

The role of electricity as a technological force in American life is fairly well known. However, one aspect of that role—the use of electric traction by railroads—has received scant attention from historians. This oversight is not surprising, since only a dozen or so railroads in the United States converted portions of their lines from steam to electric operation. Yet those that did expended a great deal of time and money on this new application of technology, and much was expected from it. The utilization of electric locomotives enabled railroads to operate heavier trains at higher speeds than had been possible with steam power, while simultaneously eliminating the need to maintain the elaborate maintenance facilities that steam locomotives required. These and a host of other technological, economic, and social advantages, all apparent by the time World War I began, seemed to assure a bright future for railroad electrification.

Ironically, in spite of having fulfilled expectations in nearly every instance in which it was applied, electric traction never gained widespread acceptance by steam railroads in this country. By 1938, these roads were operating only 3,000 electrified route-miles, a figure that represented a mere 1.2% of total route-mileage in the United States.

The onset of the Great Depression of the 1930s undoubtedly caused some lines that had been seriously considering launching major electrification programs to cancel all such plans. The conversion from steam to electric traction, a costly proposition even in the best of times, was simply beyond the financial reach of most roads in times of economic hardship. The advent of the diesel-electric locomotive on the eve of World War II, combined with the railroads' inability to recoup the prosperity of the predepression era, further undermined the allure of electric traction. The diesel offered many of the same advantages as the "pure" electric but at a much lower first cost. This was an extremely important factor to roads whose share of freight and especially of passenger traffic was steadily shrinking.

Diesel locomotives now are firmly entrenched as the principal motive power type, and the railroad industry still has not recovered its economic health. Thus many roads are once again exploring the merits of electric traction, for it continues to offer certain economies of operation (particularly as the price of diesel fuel rises) that cannot be obtained with any other form of motive power. Electric traction has retained its potential as an effective tool which railroads can use to curb the inroads made by nonrail modes of transport.

In addition, the federal as well as various state and local governments have shown much interest in electrification in recent years, for reasons that transcend corporate profit and loss statements. Electric locomotives, unlike their diesel counterparts, have the ability to draw upon (indirectly) several plentiful domestic energy sources—coal, water, and nuclear—rather than a single scarce, foreign-based one—petroleum. Consequently, electrification should prove more suitable in meeting the requirements of a national energy utilization plan. Moreover, because of certain technical characteristics, electric traction offers a sound mass-transit alternative in today's automobile-choked metropolitan areas and the corridors that connect them. In view of electrification's ability to increase the effectiveness of public energy and transportation policies, government aid to help offset the high cost of installing an electrified system appears to be a likely possibility. Indeed, public funds are already being spent to modernize and lengthen the northeast corridor electrification, between Boston and Washington.

The time seems appropriate, then, to review the experiences of the owner of the nation's most extensive electrified system, the Pennsylvania Railroad. Between 1933 and the end of its corporate life in 1968, the PRR operated more electrified track-miles (peaking at over 2,100) than any other railroad in North America. From the standpoint of both freight and passenger traffic, the PRR's electrified lines were among the busiest in the world. By these and almost every other standard, the Pennsylvania's electrification ranked as the most important yet achieved by an American railroad. Even prior to its decision in 1928 to undertake a costly long-distance electrification project, the PRR for many years had been one of the railway industry's most prominent users of electric traction.

The history of the evolution of electric traction on the Pennsylvania Railroad encompasses more than just the story of how the road implemented a change from steam to electric motive power on some of its routes. It is also a study of how a great corporation attempted to manage an important element of technological risk. The PRR tried to utilize the most efficient technology possible to combat the threats from its competitors; yet it strove mightily at the same time to keep technological risks (and therefore the economic risks that accompanied the introduction of new technology) to an absolute minimum. In other words, the road wished to follow a conservative policy with regard to the introduction of new technology—in this case electric traction—but not one so cautious that it would compromise the company's established goals of providing better service and increasing profitability. The manner in which the Pennsylvania applied electrification also illuminates the capabilities and limitations of technological innovation as a tool of corporate management.

A survey as necessarily narrow in its outlook as this one cannot fully scrutinize the interaction between the PRR's various electrification projects and those of other roads. The Pennsylvania, nonetheless, did not electrify in an economic or technological vacuum. An exploration of its use of electric traction will thus lead to at least a partial understanding of the rise (and subsequent decline) of the popularity of electric motive power on American railroads in general.

By the time the Pennsylvania Railroad undertook its first tests with electric traction in 1895, electric propulsion was no longer a novelty among railway men. Scientists and inventors on both sides of the Atlantic had been experimenting with practical applications of electricity throughout the nineteenth century. Next to illumination, probably no other application of electricity received as much attention as transportation, or more specifically, rail transportation. Until the 1870s, batteries offered the only source of power for these experiments, since neither the electric generator nor a satisfactory means of current distribution had been perfected. A locomotive powered by on-board batteries had run in brief trials on the Baltimore and Ohio Railroad near Washington, D.C., as early as 1854. Unfortunately, storage batteries themselves had not yet outgrown their infancy. They were entirely too large, their output too anemic, and their condition too fragile for railroad service. None of these primitive attempts to develop a satisfactory method of electric motive power met with more than fleeting success.[1]

With the advent of a reliable generator and an efficient means of distributing the current it produced, the cumbersome batteries were discarded. The 1870s and 1880s witnessed a new wave of experiments being performed in the United States and in Europe using locomotives equipped with small direct current (d.c.) traction motors. Thomas Edison, to single out only the most famous of many capable inventors, by 1880 had an electric locomotive pulling two cars along a short stretch of track at his Menlo Park laboratory.

The interest in electricity as a motive power source centered chiefly around converting elevated and street railways from animal and steam traction. As cities grew, their inhabitants were becoming less tolerant of the inadequacies of urban transit. Steam locomotives operating on crowded thoroughfares had caused many accidents. Horse-drawn cars, on the other hand, crept along much too leisurely to satisfy the demands of a population increasingly in a hurry. And even in an age that was seemingly oblivious to pollution, a ground swell of protest had formed in opposition to fouling of the environment by both horses and steam engines.[2]

In 1888, Frank J. Sprague completed electrification of the Richmond (Virginia) Union Passenger Railway, the first railroad electrification in the United States that was a commercial and technical success. Sprague, a Naval Academy graduate, recognized the great potential of the electric motor even before he left the Navy in 1883 to work for Edison. When Edison expressed only lukewarm enthusiasm for the use of electric traction in urban areas, however, Sprague left to seek financial backing for a scheme to electrify elevated steam railroads in New York City. Failing in that attempt, he next turned his attention to street railways and in a short time won the Richmond contract. By the fall of 1888, Sprague had 40 electric trolley cars running smoothly and reliably over 12 miles of track.[3] This was the breakthrough the electrical industry had been awaiting. Investment capital poured in. Electric street railways blossomed in city after city. By 1900, trolley cars were as much a fixture of the urban scene as boss politics and immigrant ghettos.

The almost overnight success of the streetcar lines gave rise to an abundance of longer interurban routes. Electrically powered interurban cars crisscrossed the suburban and rural landscapes in astonishing numbers as the century closed. These cars were somewhat larger and sturdier than city trolleys. They were also enclosed, whereas many streetcars remained open to the elements. Most of the interurbans, in fact, differed in few respects from the ordinary coaches used by steam railroads. Both street and interurban railways catered to the local passenger trade almost exclusively. With few exceptions, very little freight was carried, and runs were relatively short. By the same token, in most instances neither the interurbans' rolling stock nor their rights-of-way were engineered to the heavier, more exacting standards of steam railroads. By 1895, some 850 electric street and interurban lines were operating about 23,000 cars over approximately 9,000 miles of track. The companies represented a total investment in excess of $400 million.[4] Few other industries could boast of achieving such rapid growth in so short a period of time.

Interurban lines possessed several important advantages over the competitive operations of steam railways. The use of electric motors resulted in both cleaner accommodations for the traveling public and faster schedules. The electric cars had a much more rapid rate of acceleration than their steam counterparts, a feature of paramount importance on most short-haul routes, with their countless station stops. The cars' simplicity of design and construction also resulted in lower operating costs and superior dependability, as compared to steam equipment. Interurban railways, nevertheless, posed only a limited economic threat to steam railroads, which maintained a secure hold on practically all long-distance passenger and freight service. Only in and around metropolitan areas did the interurbans seriously challenge the supremacy of steam roads. There they competed mostly for commuter and other types of local passenger traffic.[5]

Beginning in the 1890s, a number of steam railroads initiated inquiries into the field of electric traction. Whether or not these roads sensed a serious threat from their competitors, they could hardly ignore the economies of operation that electricity had bestowed on the street and interurban railways. Just before the turn of the century, several steam roads went so far as to convert small segments of their lines to electric operation on an experimental basis. In 1895, for example, the Pennsylvania Railroad began electrically powered passenger service on its 7.2-mile Burlington and Mount Holly Branch. That same year, the New York, New Haven and Hartford Railroad applied electric traction in a similar fashion on its 7.1-mile Nantasket Beach (Massachusetts) line.[6] Since both of these roads based their electrifications on the technology and operating practices of the interurbans, the installations had a minimal impact on the future development of electric traction. Far more influential in molding the kind of long-distance, heavy-duty electrification that eventually appeared on the Pennsylvania was the Howard Street tunnel electrification of the Baltimore and Ohio Railroad.

The B&O and the city of Baltimore had flourished jointly since the railroad's chartering in 1827. By the 1880s, however, the B&O found itself at a very serious disadvantage to relative newcomer PRR with regard to a rail route through Baltimore. Whereas the Pennsylvania owned a direct, all-rail route through the city, the B&O was forced to halt its trains on either side of the Patapsco River and ferry the cars across. The railroad had not considered this procedure to be unduly burdensome until about 1886, when it obtained a direct route (via its own rails and trackage rights over the Reading and the Central of New Jersey) to the New York metropolitan area. The time and expense involved in ferrying cars at Baltimore severely handicapped the B&O in its effort to compete with the powerful PRR for traffic to and from New York and New Jersey.

The B&O considered a number of proposals for doing away with the car floats and establishing an all-rail route of its own. By 1890, it had decided to construct a 7-mile belt line through Baltimore, beginning near Camden Station and running north and east to Waverly, where the belt line would rejoin the main tracks to Philadelphia. Plans called for boring a 7,000-foot tunnel from Camden Station north under Howard Street. While this tunnel solved one problem for the railroad, it presented a major difficulty in its own right. The persistent 0.9% grade northbound through the bore would evoke a tremendous amount of smoke and gas from steam locomotives as they struggled up the slope. The B&O pondered the problem even as the tunnel was being built and finally concluded that the subterranean passage could not be ventilated adequately. In the spring of 1892, the railroad awarded a contract to the recently organized General Electric Company for electrification of that portion of the belt line that ran through the Howard Street tunnel and a short distance beyond. A total of 3.6 route-miles was to be energized.

On July 1, 1895, the electrified part of the new line was opened for regular service. General Electric had supplied a 96-ton locomotive comprised of two semi-permanently coupled units. A 360-horsepower motor was mounted on each of the locomotive's four axles. A single pantograph collected 600-volt direct current from a peculiar system of inverted iron troughs rather than from an overhead contact wire or ground-level third rail common to interurban and street railways. Despite some initial teething problems, the B&O's electrification soon proved to be an unqualified success. The new locomotives (GE delivered two more before the year was out) coupled onto northbound trains at Camden Station and hauled them, steam engines and all, up the grade and through the tunnel with ease. Trains weighing as much as 1,200 tons could be pulled at a steady 20 miles per hour with hardly a strain. The Baltimore and Ohio's installation, short though it may have been, marked the first time anywhere in the world that electricity had supplanted steam as a motive power source in main-line railroad service.[7]

In the perpetual darkness beneath Howard Street, General Electric had conclusively demonstrated that electric traction could master the heavy tonnages and rigorous demands of heavy-duty railroading. No longer did electric traction have to be confined to the lightly ballasted roadbed and pastoral surroundings of the interurban. Ironically, the Pennsylvania Railroad recognized the significance of this achievement to a greater degree than did even the Baltimore and Ohio. A few years hence, the PRR would undertake a far more ambitious electrification scheme of its own.

That the Pennsylvania would so quickly adopt the use of electric motive power was very much in character for a railroad that had been in the forefront of technological improvement within the railway industry for half a century. The PRR enjoyed the distinction, for example, of being the first major carrier to use steel rails in place of the more brittle iron ones. It was the first road to opt for widespread use of George Westinghouse's new air brake. The PRR's preeminent mechanical staff early on won a reputation for pioneering in motive power design, adding literally hundreds of improvements to the endless ranks of iron horses that emerged from the road's erecting shops at Altoona, Pennsylvania[8]

Furthermore, the Pennsylvania had the economic resources to sustain its taste for technological progress. By the end of the nineteenth century, the Pennsylvania and its subsidiaries sprawled over 13 states and the District of Columbia. In these states were situated some of the greatest industrial areas and most bountiful agricultural lands in the world. From its corporate headquarters in Philadelphia's Broad Street Station, the PRR in 1900 operated over 3,600 route-miles, extending from the crowded avenues of New York and Philadelphia on the east to the gateway cities of Chicago and St. Louis on the west. Into its 50,000 freight and express cars went everything from coal to livestock to the United States mail. Its crack limiteds and their crews pampered travelers with unsurpassed opulence and the most attentive service imaginable.

Nor did the Pennsylvania allow stock watering and other kinds of unsound financial manipulations—the plague of many railroads—to rob it of its vigor. Solid, no-nonsense businessmen had guided the road from the start. The honest capitalism of these men stood in sharp contrast to the unscrupulous machinations of the Goulds and Vanderbilts and Fisks of other lines. This conscientious stance by management yielded rewards of the most positive sort. By the turn of the century, the Pennsylvania was the largest, wealthiest, and busiest railroad in the country. Not without some justification had it begun to refer to itself as the "Standard Railroad of the World."[9]

Pennsylvania Railroad and connections. (PRR)

1

....

Tunnels to Manhattan

The post-Civil War years were a time of unprecedented expansion for the Pennsylvania Railroad. In 1869, the PRR gained unhindered access to Chicago by securing full control of the Pittsburgh, Fort Wayne, and Chicago Railroad. At about the same time, it won control over a direct route to St. Louis and the Mississippi River. A multitude of lesser lines were absorbed into the system during this period as well. The railroad had become so large by 1870 that a separate subsidiary, the Pennsylvania Company, was formed to operate all lines west of Pittsburgh and Erie. This was the beginning of the famous Lines West and Lines East arrangement, a dichotomy that lasted until after World War I.[1]

While the mighty Pennsylvania boasted of having pushed its steel tentacles into some of the nation's most populous cities, it could not make that claim with regard to New York City. Throughout the last years of the nineteenth century, the PRR struggled in vain to conquer the great natural barrier—the Hudson River—which lay between it and America's largest metropolis. The railroad was torn between building a bridge or boring a tunnel. Financial problems precluded the construction of a bridge. The impossibility of running steam-powered trains for long distances under water prohibited the construction of a tunnel. Then, just as the new century dawned, the Pennsylvania seized upon a technological innovation to resolve its dilemma. The electric locomotive, in making possible the safe operation of trains through long tunnels, finally allowed the PRR to achieve an all-rail entrance to the island of Manhattan.

The Pennsylvania had terminated on the west bank of the Hudson River since 1871. In that year, the railroad acquired the United Railroads and Canal Companies of New Jersey, whose main line connected Jersey City and Trenton. Passengers aboard PRR trains bound for New York City and points east had to disembark at Exchange Place, on the Jersey City waterfront. From there they were forced to take the ferry across the Hudson to lower Manhattan. If they were headed to midtown or upper Manhattan, a cab or trolley ride through the narrow, teeming streets of the city was also in order. The ferry trip alone required between 15 minutes and an hour, depending on the weather. Travelers crossing New York Bay for Long Island destinations could expect to be waterborne much longer. From the beginning, the Pennsylvania regarded this tortuous mode of transit as temporary. How could it continue to call itself the "Standard Railroad of the World" and yet be virtually shut off from the nation's largest city?

The management of the PRR was not the first to express interest in finding a better means of connecting Manhattan and New Jersey. People had been talking of a bridge over or a tunnel under the Hudson River for several decades. Not long after the PRR's purchase of the United Railroads, a former Union army colonel named Dewitt Clinton Haskin organized a company to drive a railroad tunnel beneath the river. Construction began in 1874. The Delaware, Lackawanna, and Western Railroad, which stood to lose a great deal of its ferry business if Haskin's tunnel were completed, quickly obtained an injunction against the project. Not until 1879 did the courts permit Haskin's company to resume work. By the following year, nearly 1,200 feet of brick-lined passageway had been built from the Jersey side eastward under the rushing waters of the Hudson. On July 21, 1880, a "blowout" claimed the lives of 20 tunnel workers. Investors and the general public lost confidence in the scheme. Construction came to a halt soon after the disaster, and the plan was temporarily abandoned.[2]

Although the Pennsylvania Railroad kept itself well posted on the Haskin venture from beginning to end, it held aloof from any involvement.[3] Even if the tunnel were bored successfully, it would be practically useless. The only railway motive power then available was the steam locomotive, which would promptly asphyxiate passengers and crew alike if used in a longer underwater passage. Perhaps special cable-drawn trains could be used, but this would entail construction of special cars. Transfer from steam trains to tunnel trains would have little advantage over the transfer from steam trains to ferryboats.

Several years after the extinction of Haskin's Hudson Tunnel Railroad Company, the PRR took up the matter of a Hudson crossing. It wanted nothing to do with tunnels. Instead President George B. Roberts instructed his principal assistant engineer, Samuel Rea, to "find the best bridge engineer you can hire" to advise the Pennsylvania on how it might go over the Hudson rather than under it.[4]

Rea did not have to search far. He selected Gustav Lindenthal, a 34-year-old Austrian-born, German-educated civil engineer. The two men had met several years earlier, when Lindenthal was building bridges at Pittsburgh, and Rea was stationed there as a PRR assistant engineer. Even at this early stage in his career, Lindenthal had won a reputation as one of America's premier bridge builders. For example, in his Smithfield Street bridge across the Monongahela River, completed in 1882, he had pioneered in the use of steel in place of the usual cast iron.[5]

Rea judged his friend Lindenthal to be just the man Roberts was looking for. In 1884, the Pennsylvania commissioned the Austrian engineer to work up preliminary plans for a Hudson bridge having a Manhattan terminus at Canal and Desbrosses streets. Later that year, Lindenthal laid before the road's board of directors a proposal for an immense single-span suspension bridge. Despite the enormous size of the structure, the railroad expressed considerable enthusiasm for it. From the PRR's Philadelphia headquarters, the drawings were sent to Washington for examination by the War Department. The department objected strongly to the bridge, primarily because Lindenthal had located the piers within the river's navigable channel. This unforeseen difficulty, coupled with a short but severe recession in 1884–85, caused the Pennsylvania to withdraw its support of the plan.[6]

Lindenthal refused to give up. He revised his plans to conform to the War Department's specifications and then waited for the return of more prosperous times. In 1889, on his own initiative, he persuaded the House of Representatives' Committee on Commerce to hold hearings on the matter. He wanted Congress to issue a charter to his newly formed North River Bridge Company and thereby give national sanction to the work of crossing the Hudson. Lindenthal was not alone in the campaign to have the bridge built. The incorporators of the North River firm included Rea (who was now working for the Baltimore and Ohio), financier Thomas Fortune Ryan, and future United States Vice-President Garret Hobart. The Pennsylvania Railroad took no part in the hearings.[7]

In testimony before the committee, Lindenthal cautioned that his bridge was "unprecedented in magnitude, and should not be mistaken for the routine work of an ordinary Missouri or Ohio River Bridge." Its clear center span of 2,850 feet was much longer than that of any existing bridge, suspension or otherwise. Distance between anchorages totaled 5,850 feet. Only Scotland's Forth Bridge, then being built, rivaled Lindenthal's structure in total length; but the cantilevered Forth Bridge had two clear spans of only 1,700 feet each and carried only two railroad tracks. Lindenthal suggested a total of ten tracks for his. Othmar Ammann, designer of the celebrated George Washington Bridge, 50 years later acclaimed Lindenthal's proposal as "a remarkably bold and well-conceived plan," ahead of its time in size, but sound from a structural standpoint.[8]

The Committee on Commerce listened to approving comments on the venture from many engineers, War Department personnel, and civic leaders. No one voiced any serious reservations. In March 1890, the committee reported favorably on the bill to charter the company.

It recommended that Congress as a whole take positive action, noting that

> the necessity for such improvement as the proposed bridge will greatly increase from year to year, according to the growth of the country and the increase of traffic to and from New York City, so that the speedy construction and completion of such a bridge . . . is of great importance, not only to the neighborhood of New York City, but to the country south, west, and north of it.[9]

Congress agreed. It authorized a charter for the North River Bridge Company, and President Harrison signed the document in July 1891.[10]

The lack of participation in this affair by the Pennsylvania Railroad did not signify any loss of interest in attaining entrance to Manhattan. On the contrary, the railroad felt the inferiority of its position more acutely than ever. The New York Central, with its water-level route down the Hudson Valley terminating in upper Manhattan, enjoyed a clear competitive edge over the PRR in attracting east-west passenger traffic. Nevertheless, railroad planners predicted a steady growth in the Pennsylvania's overall passenger business to and from the New York area. They also forecast a continuing rise in local passenger traffic, in light of the seemingly unrestrained expansion of New York City and its environs. The PRR estimated that by 1900 it would be moving over 33 million people a year to and from Manhattan.[11] To believe that ferries could adequately handle this traffic was unrealistic. The time was rapidly approaching when the railroad would have to confront its problem decisively, one way or another.

The Pennsylvania offered to participate in Lindenthal's grand design, but it refused to bear the financial burden by itself. Because of the tremendous cost of constructing the bridge and its approaches (estimated to be about $100 million), the railroad envisioned the project as a joint venture of all the roads terminating at the Jersey waterfront. After the North River Bridge Company had been chartered, representatives of the PRR approached the half dozen or so other rail lines which would most likely be involved. At that time, the Hudson ferries were transporting approximately 50 million people annually back and forth between New York and New Jersey. Slightly over half of these passengers were en route to or from the Pennsylvania's trains. Most of the remaining business belonged to the Erie, the Lackawanna, and the Central of New Jersey-Reading-B&O combine. Would not these other roads, reasoned the PRR, also be interested in a bridge to Manhattan? As events unfolded, they would not. The smaller companies hesitated to put forth a large capital outlay only to have their giant competitor reap most of the eventual benefits. The ferry problem was not so severe for railroads having a much smaller business, nor did it produce the humiliation suffered by the proud Pennsylvania. Besides, many railroad men remained unconvinced of the practicality of such a daringly long suspension bridge as Lindenthal had suggested.[12]

In the summer of 1892, President Roberts dispatched Samuel Rea, his new assistant in charge of construction and purchases, to England to inspect the operation of the new City and South London electric subway. Upon his return, Rea was to prepare a report on the feasibility of the various bridge and tunnel ideas that had been talked about for the New York area.

Rea was an ideal choice for such an assignment. Born in the canal and railroad town of Hollidaysburg, Pennsylvania, in 1855, he had commenced his railroad career at the age of 17 when he took a job as chainman on a PRR engineering gang. Later he left the Pennsylvania to take engineering positions on other railroad and bridge construction projects. He returned to the Pennsylvania in 1879 as principal assistant engineer. Ten years later, he again took his leave, this time to become chief engineer for the new Baltimore and Ohio Belt Railroad. In his new position, Rea had the responsibility of designing and overseeing the construction of the Howard Street tunnel. It was at this time that Lindenthal persuaded him to become an incorporator of the North River Bridge Company. Poor health forced Rea to resign his Baltimore job in 1892, before the B&O had committed itself to the use of electric traction. He then returned to the PRR as assistant to President Roberts.[13]

While in London, Rea observed with interest the new electric locomotives used by the City and South London Railway. The C&SL was the world's first successful railway subway system, mainly because of the utilization of these smokeless engines. It ran beneath the streets of London for about 3 miles and connected two of that city's major stations. The success of the subway did not convince Rea that electric traction could handle the much heavier tonnages found on American steam railroads. In his report to Roberts and the Pennsylvania's directors, he argued against building a tunnel. Instead he urged that the railroad return to the Lindenthal plan for a suspension bridge.[14]

Samuel Rea delivered his report in 1893, two years before General Electric and the Baltimore and Ohio combined to make history under the streets of Baltimore. The electric locomotives which he had seen in the London underground weighed but 10 tons apiece, making them much too light for the tasks the PRR had in mind. Therefore, Rea had little enthusiasm for a tunnel scheme whose entire operation was predicated on the use of electric traction. His recommendation of a bridge was understandable, but it was also ill timed. A financial panic occurred early in 1893, plunging the nation into one of the most severe periods of economic depression it had ever experienced. Whether or not the Pennsylvania could convince its competitors to cooperate in building a bridge was now a moot issue. Lindenthal's ill-starred proposition collapsed along with the earnings of its potential backers.[15]

Supporters of the North River Bridge Company knew that good times were bound to return. In the interim, perhaps something might be done to allay the fears of those who doubted the strength and safety of such a long bridge. They prevailed upon Congress to create an official commission of engineers to investigate and pass judgment on the matter. The North River's backers were also acting with an eye toward countering the claims of a rival franchise, the New York and New Jersey Bridge Company, which had received a Congressional charter shortly after the North River organization. Its proponents deprecated the idea of a suspension bridge and were aggressively advancing plans for a colossal cantilever.[16] In June 1894, Congress directed President Grover Cleveland to appoint a five-man board of engineers to determine if Lindenthal had overreached himself in calling for a span of 2,850 feet. Cleveland chose Major Charles W. Raymond, the Army Corps of Engineers officer in charge of defenses and harbor improvements at Philadelphia, to head the panel. The other members were private civil engineers.

Several months later, the board presented its findings. Lindenthal's calculations were sound. Indeed, Major Raymond and his associates said that either a suspension or a cantilever structure having a clear span of up to 3,100 feet would pose little risk. The board refused to show preference for either type.[17]

The project lay dormant for several years as the Pennsylvania Railroad and the country in general struggled to regain financial health. In 1899, the PRR reported a net income of $11.2 million, up from a depression-era low of $9.3 million in 1894. For 1900, net income had climbed to $18.2 million.[18] By this time, the management of the Pennsylvania believed the company had recovered sufficiently to explore once again the possibilities of a bridge or tunnel to Manhattan. George Roberts no longer presided over the destiny of the railroad, however. He had succumbed to a heart ailment in 1897 while in the seventeenth year of his presidency. Frank Thompson was elected to succeed Roberts but served only two and a half years until he, too, died. In June 1899, the company's board of directors selected one of their own members, Alexander J. Cassatt, to succeed Thomson.

Cassatt, a native of Pittsburgh, was 59 years old. His wealthy family was able to provide him with an excellent education, first at Germany's Darmstadt University and then at New York's Rensselaer Polytechnic Institute, from which he graduated with a civil engineering degree. Cassatt served with the PRR in a number of engineering capacities, and then, as was the custom for its ablest engineers, entered the ranks of management. By 1874, he had become the road's third vice-president. A crisis in his career occurred several years later. Cassatt had been mentioned as an outstanding prospect for PRR president for a number of years. When that post fell vacant in 1880, however, the board of directors elected Roberts. Cassatt retired shortly thereafter, amid rumors that he was miffed at not being offered the top spot. He spent the next 17 years in semiseclusion, retaining only a seat as a PRR director.[19]

Cassatt also retained an interest in the railroad's attempts to enter New York. Indirectly, he was responsible for Roberts's dispatching Rea to England. In 1891, Cassatt had persuaded Roberts to have the railroad's engineers investigate the feasibility of a tunnel between Staten Island and Long Island. The tunnel was part of a scheme Cassatt had developed whereby the PRR would not cross the Hudson at all. Instead Cassatt suggested that the Pennsylvania extend its main line along a more southerly course, from New Jersey via Staten Island to Long Island. At Long Island, the line would divide, one fork entering Manhattan over an East River bridge, the other continuing north to a direct connection with the New York, New Haven, and Hartford Railroad. Such a plan would give the railroad both an all-rail entry into Manhattan and a valuable through route for traffic to and from New England. A PRR engineering study completed late in 1891 showed that Cassatt's proposal was feasible, although the tunnel might not be able to accommodate full-size equipment. This left Roberts and the board of directors in a quandary. Should the railroad press for a Hudson bridge, a Hudson tunnel, or Cassatt's bridge-and-tunnel idea? It was at this point that President Roberts asked Rea to visit London to report on the tunnel railroad there.[20]

Cassatt brought to his new post as chief executive a determination to act. Proof of this came during the spring of 1900. The road's directors authorized the purchase of 135,851 shares of stock in the Long Island Rail Road at a cost of $6.8 million. The transaction placed under Pennsylvania ownership 56.6% of the Long Island's capital stock, thereby giving the PRR a controlling interest in the smaller line.[21] The Pennsylvania would hardly have closed such a deal unless it contemplated construction of a direct line between the two lines in the near future. President Cassatt told his company's stockholders in May, for example, that the railroad planned "to use a part of the Long Island Rail Road to form a more convenient connection between your system and the New York, New Haven and

Hartford Railroad Company."[22] He did not elaborate further on this "convenient connection," but for passenger traffic it could only have been a direct physical union of some kind.

The Long Island had received its charter from the state of New York in 1834, ranking it among the oldest common carriers in America at the time of its acquisition by the Pennsylvania. Its original incorporators had hoped the road would become part of a rail-steamship line between New York and Boston. Their dreams were dashed when predecessors of the New Haven completed an all-rail line through Connecticut and siphoned off most of the Long Island's through traffic. In 1850, the Long Island went into receivership. Under new management, it reorganized. strictly as a carrier of local passengers and freight. The rapid expansion of New York and the corresponding growth of suburban and resort communities on Long Island gave the railroad plenty of short-haul business.[23]

During the 1890s, the Long Island had several feasibility studies done in regard to the driving of tunnels under the East River.[24] Ferries could barely cope with the burgeoning number of commuters. A better means of transit had to be found with a minimum of delay. At the same time, Brooklyn city officials, prompted by complaints from local residents, were pressuring the Long Island to do something about the choking clouds of smoke issuing from its hundreds of coal-burning locomotives. Trains sometimes ran at rush hour intervals of only two or three minutes on key routes. The incredibly congested terminal areas where riders changed from coach to ferry compounded the problem. Add to this the crowded, residential nature of western Long Island, and the railroad faced a smoke-abatement problem of the first order. After a series of negotiations with community leaders, the Long Island tentatively agreed to bury a number of its most heavily traveled tracks below street level, if the city of Brooklyn would pay part of the construction costs. The program as a whole was popularly known as the "Atlantic Avenue improvements," centering on the Flatbush terminal.[25]

Two years later, when the PRR bought into the Long Island, railroad and public representatives had expended an inordinate amount of verbiage on the improvements but very little cash. Brooklyn remained hidden under a cloud of coal smoke, and ferries filled to capacity with commuters continued to ply the East River.

In the early days of 1900, as the Pennsylvania's directors debated the purchase of Long Island stock, they also heard reports on the likelihood of extending the PRR's own line into Manhattan. Again the road's engineers favored a bridge. So once more Cassatt and his lieutenants hauled out Lindenthal's blueprints and made the rounds of the other railroads in search of financial aid. The smaller lines still refused their operation.[26] The Pennsylvania decided to press the issue no further. Yet feelings of frustration and dejection were absent from the Broad Street offices. The PRR had concluded even before broaching the bridge proposal another time that it was no longer the sole alternative. A series of tunnels could probably be bored after all. The railroad's previous opposition to subaqueous tunnels had always rested on the fact that steam locomotives simply were not suitable for service in them. A corollary to this view assumed that no other type of power could satisfactorily replace steam. Hence every tunnel scheme for New York that was ever put before the board of directors until now had in effect been rejected before it could even be heard.[27]

Although Samuel Rea was temporarily absent from the PRR management in 1890, he summed up its view very well that year in testimony before the House of Representatives' Committee on Commerce. Speaking in support of Lindenthal's bridge, Rea admitted that tunnels might be another solution, "but they are not in favor either with the railroads or with railroad passengers. They are expensive to maintain, and disagreeable to the passengers, who would probably prefer crossing on ferries to trusting themselves on slow trains running through damp and chilly submarine tunnels."[28]

By 1900, however, the B&O had completed five years of successful operation with electricity in its Howard Street tunnel. Now the once-skeptical Rea had become the PRR's most zealous advocate of the tunnel approach. He had also become fourth vice-president and was in a position to take his argument directly to Cassatt. Cassatt, although a capable engineer in his own right, was not entirely won over. Coincidentally, the Orleans Railway extension in Paris was nearing completion at this time. The French line included a 3-mile underground segment through which trains were to be pulled by electric locomotives. Rea advised Cassatt to be on hand for the opening ceremonies and observe for himself the potential of electric traction. The PRR president made the journey. Upon his return to Philadelphia, he was as convinced as Rea of the practicality of electric motive power for use under the Hudson.[29]

In mid-December 1901, President Cassatt publicly confirmed his railroad's decision to secure entrance to Manhattan via a series of tunnels. In his statement to the press, Cassatt said the PRR was certain that the use of tunnels and electrically powered trains was by far the most practical as well as the most economical solution. The tunnels would be in the best interest of the railroad and the city of New York and surrounding areas. Lindenthal's bridge, having a deck 150 feet above the river, would have required very long approaches at both ends so that the trains could ascend the level of the deck without facing an unduly severe grade. Even then, Lindenthal planned to have trains climbing a steep 1.4% slope on both the New York and

New Jersey sides. The resultant locomotive pyrotechnics would have aggravated the smoke problem in the area, and the bridge would have consumed a large quantity of choice property while depressing the residential value of land which lay anywhere near it and its grimy tenants. Cassatt promised that the tunnels "would not be objectionable in any way." There would be no smoke, no noise, no dirt, and, perhaps best of all, surface properties would be left almost untouched. Real estate could only rise in worth.[30]

Well before the PRR president had disclosed his road's intentions, agents for the railroad had purchased practically all the land it would need in New Jersey, Manhattan, and Long Island. The Pennsylvania made strenuous efforts to keep these moves private, lest speculators be aroused and drive the already high land prices even higher.[31] Securing the property was only one of several preliminary actions which had to be taken. The railroad also had to obtain approval of its plans by the various governmental jurisdictions involved. Cassatt and his associates turned their attention to that end next.

The PRR did not unveil the details of its New York improvement plans. These were still being formulated. In general terms, however, the railroad wanted to bore two single-track tunnels from just west of Bergen Hill on the New Jersey side eastward to Ninth Avenue in Manhattan. The western end of this extension would connect with the PRR main line near Newark. On the east, a magnificent new station would be built between Sixth and Seventh avenues and between Thirty-first and Thirty-third streets. Four single-track tunnels would leave the station eastbound, pass under the East River, and emerge in the vicinity of Long Island City, where a junction would be established with the Long Island Rail Road. PRR through trains would enter and leave through the Hudson tubes. Most locals were to continue to receive and discharge their fares at Exchange Place. Conversely, the East River passages would be utilized by the Long Island's commuter fleet, as well as by the PRR.[32]

The Pennsylvania won quick approval of the scheme from New York and New Jersey authorities. New York City's Board of Aldermen was the last governing body to give its official sanction. In October 1902, this body extended its blessing to the project on the condition that the railroad assure it that "electricity or other approved power not involving combustion" would always be used as motive power. The Board of Aldermen wanted to take no chances that lives might be lost or that smoke-belching steam locomotives would one day foul the air of mid-Manhattan. Furthermore, the railroad had to agree that it would not disrupt traffic on the streets above and that it would make good on any property damage.[33]

Because of the enormous cost of these improvements, and because

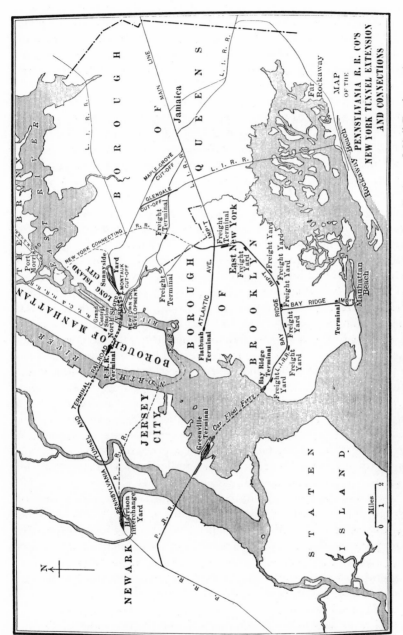

New York Tunnel Extension of the Pennsylvania Railroad. (American Society of Civil Engineers)

they included some novel engineering features in addition to the use of electric traction, the Pennsylvania's board of directors desired to have a panel of outside experts review all plans before so much as a shovelful of earth was turned. In January 1902, the directors moved to create an impartial board of engineers to act as consultants. Heading the group as chairman was Charles Raymond, now a colonel of engineers. Sitting as members were Gustav Lindenthal, Alfred Noble, William H. Brown, and Charles M. Jacobs.[34]

Lindenthal, of course, had been working closely with the PRR for 15 years. His consent to sit on the engineering board in no way signaled a loss of faith in his great bridge. For the rest of his life (he died in 1935), he lobbied unstintingly for its construction. Alfred Noble had wide experience in all kinds of bridge and canal construction. President McKinley had recently honored him with an appointment to the Isthmian Canal Commission. William Brown was chief engineer of the Pennsylvania and had overseen the drawing of the plans which the board would review. He, too, had many years of experience in the field of railway and civil engineering.[35]

Jacobs, an Englishman, was highly esteemed in engineering circles for his work in building a number of important railway tunnels. New York's very first subaqueous tunnel, a gas line connecting Manhattan and Brooklyn, had been a product of his engineering talents. It was Jacobs whom the Long Island Rail Road engaged to study the feasibility of East River tunnels and to design the Atlantic Avenue improvements. He and Cassatt first conferred on the matter of a Hudson tunnel when the PRR chief executive stopped in London on his return from the Orleans Railway extension opening. At that meeting, Jacobs stoutly defended the feasibility of a tunnel, arguing that a bridge would present many more technical difficulties in its construction.[36]

In asking Colonel Raymond to head the board, the railroad was hoping to prevent a recurrence of complications with the War Department. By all standards, Raymond was an excellent choice. He had held the Philadelphia harbor post since 1890. The board of trade of that city praised him for his "unrivaled technical skill and executive ability,"[37] no doubt a quality which Cassatt found especially attractive. Superintending the work of the PRR's board of engineers would demand a man with considerable administrative capabilities. Raymond was also familiar with the situation at New York, having chaired the 1894 inquiry into safe bridge lengths.

The Pennsylvania added a sixth member to the board in 1902. Electrification was to be the element on which the whole undertaking would rise or fall. None of the men on the board as it was originally composed had much experience in the field of traction, however. Therefore, the railroad hired George Gibbs, a consulting electrical engineer of some repute, to be the board's specialist in developing suitable motive power.[38]

The board of engineers faced a fourfold task: review the PRR's plans, modify them if need be, prepare detailed cost estimates for the entire project, and finally oversee actual construction work. The board reported directly to Vice-President Rea, who in turn would keep Cassatt and the road's directors informed. Colonel Raymond and his colleagues were not to allow construction costs to be the decisive factor in their work. As Raymond later wrote, "The management earnestly impressed upon the Board throughout the whole period of its labors that the Extension and facilities were to be designed and constructed without regard to cost as a governing factor, the main consideration being safety, durability, and proper accommodation of traffic."[39]

The PRR incorporated two separate companies in 1902 to assume the legal responsibility for implementing the New York improvements: the New Jersey Tunnel and Terminal Railroad Company and the New York Tunnel and Terminal Railroad Company. In 1907, the Pennsylvania consolidated the two into a single Pennsylvania Tunnel and Terminal Railroad Company. The parent road owned all the stock of these subsidiaries.[40]

Meanwhile, the engineering panel began its task of reviewing the railroad's original plans. It made few alterations of note, and by the end of 1903 the PRR was ready to award the construction contracts. The board had decided to divide the overall project into several autonomous segments, or divisions. Each board member would then supervise the course of the work on his own division. Alfred Noble was assigned the East River division, with its pair of tunnels. Chief Engineer Brown had charge of the Meadows division at the west end of the improvements. He was primarily responsible for building a double-track fill eastward across the Hackensack Meadows to the west portals of the Hudson tubes at Bergen Hill. As chief engineer of electric traction and terminal station construction, George Gibbs was responsible for designing the track work for the new Pennsylvania Station and overseeing the construction of the station itself. (The job of excavating the station's foundation belonged to Jacobs's North River division.) Gibbs was also to lay out the railroad's passenger car storage and maintenance yard at Sunnyside (Long Island City) and to supervise the installation of all communications and signal equipment. These duties were in addition to his primary function of directing the PRR's electric traction program. Gustav Lindenthal resigned from his post late in 1903, his talent as a bridge builder no longer being essential to the railroad.[41]

The civil engineering aspects of the PRR's New York extension were of such magnitude that they can only be summarized here.

What was most noteworthy was that before the railroad could even begin to perfect a workable system of electric transport, it had to overcome some physical barriers of the most formidable nature. To a limited extent, precedent had already been established for submarine tunnel construction. There was Jacobs's Ravenswood Gas Company tunnel under the East River, for example; but pushing gas through a tunnel was not the same as pulling a train through one. On the other side of the Atlantic, British engineers had driven several tunnels beneath the Thames River. North America's only subaqueous railroad tunnel belonged to the Grand Trunk Railway. This was a passage of some 6,100 feet under the St. Clair River, north of Detroit. The tunnel, completed in 1891, connected Port Huron, Michigan, with Sarnia, Ontario. The Grand Trunk encountered no extraordinary problems in boring the tunnel, which boded well for the PRR's own project, but operating trains through it was another matter. Despite taking precautions, such as using specially built anthracite-burning locomotives, the Grand Trunk suffered several fatal accidents in the tunnel by 1900.[42]

The Pennsylvania Railroad consigned almost all the work of its New York improvements to private firms. The John F. O'Rourke Engineering and Construction Company of New York City and S. Pearson and Son of London, England, were allotted the two most rigorous jobs: the Hudson and the East River tunnels. O'Rourke, a veteran construction firm of national repute, got the Hudson River contract. The Pearsons, having won worldwide fame a decade earlier for the erection of the Forth Bridge, took similar honors for the East River work. The railroad doled out the remaining work to a host of smaller contractors and subcontractors. It retained the architectural firm of McKim, Mead, and White to design Pennsylvania Station. Partners Charles Follen McKim and William R. Mead personally attended to the preparation of plans for this structure. Work got under way in Manhattan on the massive new station in July 1904, at about the same time as the Pearsons and O'Rourke started drilling the tunnels. Pennsylvania Station was to occupy 8 acres above ground, with space below for 21 tracks and 11 platforms, all on a single level. Before the erection of the station proper could begin, the contractors had to excavate a huge amount of earth and rock. By 1908, after nearly four years of relentless toil, over 3.5 million tons of Manhattan subsoil had been removed. The railroad barged most of this waste material over to New Jersey to be used for the Meadows fill and for the new waterfront freight yard the PRR was building at Greenville.[43]

The Meadows division comprised mostly bridge and fill work. At Harrison, New Jersey, located about a mile east of Newark, the two new Pennsylvania Tunnel and Terminal tracks veered northeastward from the PRR's main line to Exchange Place. After crossing above the main line, the two tracks traversed 6 miles of meadowlands—an idyllic euphemism for several thousand acres of mosquito-infested marshland—on a levee some 25 feet high. The line also passed over the main stems of some of the Pennsylvania's competitors at the same time, and hurdled the sluggish Hackensack River by means of a steel drawbridge. At the western face of Bergen Hill, the tracks separated, then dipped underground in preparation for their journey under the river. Just west of Bergen Hill, the railroad planned to widen the fill to accommodate a half dozen or more tracks. At this small yard, the electrically powered trains outbound from Manhattan would pause for the change to steam locomotion. Likewise, trains from the west would halt here to have their steam engines replaced by electric locomotives. The PRR also proposed to locate its own generating plant here to provide power for the entire electrified zone.[44]

The New York tunnel extension and related improvements comprised the most ambitious construction project yet undertaken by a single railroad. The Pennsylvania was pouring an average of $1 million every fortnight into the venture, a staggering amount of money for the day. Ten thousand men labored under the direction of Raymond and the other engineers during the project's peak years.[45] Newspapers such as the *New York Times* considered it a public obligation to keep their readers informed of the most trivial details of the construction. The trade journals had a field day as they proclaimed over and over that the improvements were just a foretaste of what modern technology would bring to American people in the decades ahead. Even the staid *Scientific American* seemed a little in awe of the whole thing.

In September 1906, O'Rourke's sandhogs holed through the first of the Hudson tunnels. The second followed a month later. By the spring of 1908, the Pearsons had completed boring all four East River tunnels. The project was far from being finished, however. More concrete had to be poured in the tunnels' interiors. Communications and signal equipment had to be installed. At first, the board of engineers had thought that no forced ventilation system was necessary, since passing trains would suck in fresh air and expel stale air. This was a tribute to the cleanliness of electric traction. The board eventually reconsidered the matter to take into account passengers on trains which might for some reason be stalled in the middle of the tunnel. Ventilation machinery thus had to be custom-built halfway through the job. Early in 1908, the granite shell of Penn Station had hardly begun to take shape. Construction of the Seventh Avenue subway, by which the city would plug its own transportation network into that of the PRR, had not even commenced.[46]

CROSS-SECTION OF RIVER TUNNEL

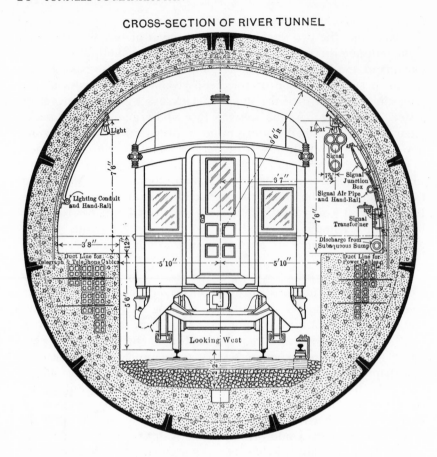

Cross section of one of the PRR's Hudson tubes. Note the third rail in the lower right corner. (American Society of Civil Engineers)

The Pennsylvania was by no means the only railroad company in the New York area building subaqueous tunnels or studying the possibilities of electrification. Three steam railroads were in the process of assigning terminal operations to electric traction, while a fourth road, just beginning its corporate life, was driving two sets of tunnels under the Hudson and had already decided to use electric motive power in them.

When D. C. Haskin's Hudson Tunnel Railroad Company had gone bankrupt in the early 1880s, work had stopped, but the notion of its eventual completion persisted. In 1888, a new company, founded on the strength of British capital, contracted with S. Pearson and Son to resume construction. Four years later, work ceased once more, this time because the British investors had no more money to put into the enterprise. The tunnel lay in a soggy state of limbo for the next decade.[47]

In 1901, William Gibbs McAdoo, a young Georgia-born lawyer who had arrived in New York only a few years earlier, revived the project. McAdoo later confessed that his guiding motive at the time was simply to make himself rich. Performing a public service by offering travelers a direct connection between New Jersey and New York he considered a secondary impulse at best. McAdoo retained the ubiquitous Charles Jacobs to examine the old Haskin bore. When the Englishman pronounced it sound in spite of years of neglect, McAdoo proceeded at once to secure enough financial backing to resurrect the tunnel for yet a third time. McAdoo had neither the engineering expertise nor the personal contacts—despite his office's Wall Street address—to win over many influential supporters.[48] What swayed such potential investors as Judge E. H. Gary of the United States Steel Corporation was not the Georgian's alluring rhetoric, but rather the earnest intonations of A. J. Cassatt. In announcing his railroad's plans in December 1901, Cassatt had stated frankly that the PRR was going to use its investment dollars to improve facilities for long-haul passenger service. Local riders to and from Manhattan must continue to endure the torture of the Hudson River ferryboats. Even a member of the road's board of engineers admitted that "the downtown local business was not provided for properly."[49] The Pennsylvania—consciously or otherwise—had set before McAdoo a golden opportunity. Why should people continue to take the ferries across the water if they could glide smoothly under it in less than a quarter of the time?

Thus came into existence the Hudson and Manhattan Railroad Company, of which McAdoo was president and chief booster. The firm's incorporators did not hesitate when the time came to decide on motive power. The road for all intents would be an interurban line, so self-propelled multiple-unit (m.u.) cars were ordered rather than locomotive-drawn equipment. (The multiple-unit method of operation was perfected by Frank Sprague in the 1890s. It was an electropneumatic system that permitted two or more electrically powered cars or locomotives, when coupled together, to be operated as one; that is, one engineman using one set of controls could operate several cars or locomotives as a single unit. Cars equipped with this type of control were thus termed "multiple-unit cars.")

Construction began during the summer of 1902. Since the Pennsylvania did not begin its own tunnel excavation until 1904, the H&M's experience in boring subaqueous passages provided the PRR's engineers with much valuable information. The Hudson and Manhattan ran its first regularly scheduled train beneath the Hudson on February 28, 1908. The line extended from the Lackawanna terminal at Hoboken eastward under the river to Sixth Avenue at Washington Square. There it turned north, running below Sixth Avenue to a station at Thirty-third Street.[50]

While the H&M's tubes were located south of the PRR's, travelers destined for or departing from the lower Manhattan business district still found them inconvenient. McAdoo longed to build another set of tunnels still further down the Hudson. They would enter Manhattan somewhere around Cortlandt or Fulton Street. On the west, he hoped that they could surface in the vicinity of the Pennsylvania's Exchange Place terminal. Before he could obtain financing for this scheme, however, he had to get the PRR to agree to it. In the first place, the majority of the passengers who would use the new tunnels would come from PRR trains terminating at Exchange Place. Secondly, the right-of-way of these "downtown" tunnels, as they came to be known, lay partly under PRR property in New Jersey.[51]

McAdoo believed that the Pennsylvania might be glad to come to an understanding of some kind with the H&M. He reasoned that the Pennsylvania's management viewed traffic between Harrison and Exchange Place as "a sort of trailing loose end of their enormous scheme." Early in 1903, he and partner Walter Oakman traveled to Philadelphia to discuss the proposal with President Cassatt. To McAdoo's great relief, Cassatt approved of the idea of downtown tubes. In fact, he was enthusiastic about it. "You are about to put our ferry out of business," he remarked to his two visitors, "but the Pennsylvania Railroad believes in providing the best facilities for its patrons, and, as your tunnels will do that, we'll hook up with you."[52]

Several years later, the PRR abandoned practically all of its downtown commuter business to the Hudson and Manhattan. The agreement of 1903 called for the Pennsylvania to run its trains to Exchange Place as usual. There, patrons would have the option of either taking a PRR ferry or an H&M tube train into Manhattan. In 1906, two years before the H&M collected its first fare, the two railroads signed a new pact. Instead of disembarking at Exchange Place, PRR passengers headed downtown could change to McAdoo cars at Harrison, 9 miles west of the Jersey City waterfront. The agreement also allowed H&M trains trackage rights over part of the PRR's New York division main line between Harrison and Exchange Place, so that McAdoo's company did not have to build a completely new line between those points. The Pennsylvania had originally intended to institute a shuttle service between Newark and its ferries for riders going to lower Manhattan. The 1906 accord killed that idea, although the PRR continued joint service with the H&M between Newark and Jersey City. Indeed, the PRR-owned ferries plied the Hudson for another 40 years, and PRR trains ran to Exchange Place even longer. But most passengers preferred to board the Hudson and Manhattan cars at Newark or Harrison. Thus the H&M became the main commuter line into lower Manhattan as the Pennsylvania concentrated its energies elsewhere.[53]

Given the expensive improvements already in progress, the PRR could not have afforded to build its own tunnels between Jersey City and downtown Manhattan. In that light, its abdication to the H&M stemmed from a realization of the facts of economic life. In addition, the Pennsylvania obviously believed that more money was to be made from long-distance traffic than from local transit, even in these days before the automobile. From that standpoint, McAdoo was perhaps more nearly correct than he suspected in describing the local traffic to New York as a "trailing loose end."

The agreement of 1906 ultimately had the effect of expanding the Pennsylvania's electrified zone. Since many of its trains would have to stop at Harrison to permit riders to switch between its own and the H&M's cars, the PRR decided to make that the spot where electric and steam locomotives would be exchanged. The western boundary of electrification was thus set at what the railroad at first called "Harrison Transfer," rather than at Bergen Hill.[54]

In uptown Manhattan, two other rail electrification projects were under way. Four hundred trains originated or terminated daily at

Manhattan's Pennsylvania Station in the late 1920s. (Conrail)

New York Central's Grand Central Station at Forty-second Street. Roughly half of this number belonged to the New Haven, whose trains had trackage rights over the Central beginning at Woodlawn, New York, 12 miles north of the station. The New Haven shared Grand Central's facilities as well. Cornelius Vanderbilt had built Grand Central in 1871. By the turn of the century, it was much too small to accommodate such a vast amount of business. Inasmuch as the New York Central expected traffic to increase even more, it concluded that a new station would be necessary. With the price of land in upper Manhattan rising as fast as the new skyscrapers in the lower part of town, the railroad planned to expand its facilities below ground rather than above. Subterranean operations of any real scope had to be based upon the use of electric traction. Even using the present antiquated station, steam trains had to run the gauntlet of the smoke-filled Park Avenue tunnel (actually a partially covered cut 2 miles in length) north of Grand Central. In addition, electrification would eliminate the dreadful smoke problem which plagued those New Yorkers who were unlucky enough to live or work near the railroad's main line. The New York Central had been studying electrification as early as 1899. As of January 8, 1902, it had not reached a decision one way or the other. Early that day, the engineman of an inbound Central local failed to spot a red signal hidden in the gloom of the Park Avenue tunnel. Seconds later, his train slammed into the rear of a stalled New Haven commuter train. Seventeen people died.[55]

That tragedy determined the course of electrification on both railroads. The public was aghast at such a calamity. Shock soon turned to anger as New Yorkers petitioned their state legislature to take action to prevent a recurrence of the disaster. Bowing to citizen pressure, the legislators passed a law in 1903 which prohibited the use of steam locomotives south of the Harlem River after July 1, 1908. The New York Central and tenant New Haven had no recourse but to replace their coal burners with electric traction.[56]

The Central electrified its four-track main line as far as Croton, 34 miles north of Grand Central, and its double-track Harlem division as far as White Plains, 24 miles out. In conjunction with these improvements, the railroad finally implemented its plan to build a new terminal. Begun in 1903, the palatial structure cost $65 million and was matched in architectural magnificence only by the new Pennsylvania Station. The Central's contractors confronted construction problems of greater severity than those faced by the PRR's builders. Grand Central Terminal (as opposed to the old Grand Central Station) was not opened to the public until early in 1913. By that time, electric trains had been running in and out of the old depot for nearly seven years.[57]

The New Haven probably could have avoided electrification, since Woodlawn Junction lay north of the Harlem River. All it would have had to do was supply its own electric locomotives for the jaunt from Woodlawn into Grand Central. For a number of reasons, however, the New Haven also opted for electric traction and proceeded to electrify its multiple-track main stem between Woodlawn and Stamford, Connecticut, a distance of approximately 20 miles. It took delivery on its first locomotive in 1905. By 1907, all New Haven trains running to Grand Central Station were electrically propelled.[58]

The third steam railroad to electrify in the New York area was the Long Island. Officials of that road had discussed, planned, and studied electrification for years. Undoubtedly, it would have had to come sooner or later. State and local government leaders kept hammering at the railroad to clean up the air. At this time, the Long Island was hauling more than 100 million people annually. Density of traffic alone argued for serious consideration of electric traction.[59] When the Long Island's seat of power shifted to the boardrooms of Philadelphia, action was assured. The railroad then had the money, the prestige, and the management skills of one of the world's most dynamic business institutions.

In 1905, the Long Island began the much-debated Atlantic Avenue improvements. It constructed a new underground terminal at Flatbush Avenue. From there, passengers had easy access to bridges and ferries (and eventually subways) to Manhattan. The railroad placed several miles of main line directly beneath Atlantic Avenue. Formerly the tracks had run right down the middle of that thoroughfare. From Flatbush, the electrified segment stretched eastward as far as Belmont Park in Queens and southward across Jamaica Bay to Far Rockaway. This first stage of the Long Island's electrification totaled 44 route-miles and 110 track-miles, making it the largest steam railroad conversion in the United States up to that time. Of greater consequence was the fact that the Long Island's switch to electric traction signified the first time ever that an American railroad had electrified a sizable portion of its main line.[60] Prior to 1905, all existing steam railroad electrifications except that of the Baltimore and Ohio had involved lightly built branch lines. The B&O's 3-mile piece, its historical significance notwithstanding, hardly qualified as a main-line operation of any real distance.

Cleaner and faster service attracted an ever increasing number of riders. The Long Island reacted by adding several extensions to its electrified lines. Seven years after the original installation, the railroad was operating 89 route-miles and 188 track-miles. Its roster of m.u. cars (it owned no locomotives) had swollen to about 400.[61]

Up to the end of 1912, the Pennsylvania Railroad had advanced over $20 million to the Long Island in order to help defray the cost of electrification and related improvements. As large as that sum may have been, the PRR still did not consider the Long Island's commuter business to be a very worthy investment. True, in 1905 the Long Island's gross receipts from passenger traffic had increased by about one-third over what they had been in 1895; but its freight business had risen by a like amount. Net earnings from freight haulage were significantly higher than those from passenger service.[62] Furthermore, when discussing the initial purchase of Long Island stock with PRR shareholders in 1900, President Cassatt never made specific reference to the smaller road's commuter activities as an attraction for the Pennsylvania. Ten years later, the PRR issued an official booklet commemorating the completion of the New York improvements. It, too, made no mention of the Pennsylvania's eagerness to become involved with the Long Island's commuter business, although it did list a number of other reasons for the purchase of a controlling interest in the road.[63]

In all likelihood, the Pennsylvania's willingness to assist in underwriting the Long Island's electrification stemmed more from a desire to cut operating costs than from an effort to increase overall profits from commuter-type services. Cassatt had told his stockholders, however, that the PRR's New York improvements "will so increase its [the Long Island's] business as to make your investment in its shares directly profitable."[64] On January 18, 1906, Cassatt sent a letter to the Board of Rapid Transit Commissioners of the City of New York in which he outlined his railroad's comprehensive plan for the New York metropolitan area. Cassatt first recounted several goals which were well on their way to being realized: tunnels from New Jersey through Manhattan to Long Island, a monumental new station in mid-Manhattan, and partial electrification of the Long Island Rail Road. Then he interjected a couple of objectives not originally included in the general program of improvements: having the New York Connecting Railroad (already incorporated) construct a link between the New Haven's terminal at Port Morris, New York, and the PRR's own Sunnyside yard on Long Island; and utilizing Long Island Rail Road trackage as part of a through freight route between Port Morris and the Pennsylvania's Greenville, New Jersey, yard.[65]

A connection between the New Haven and the PRR at Sunnyside would allow through passenger service to be operated to and from New England via Penn Station. Without such a link, passengers would not only have to change trains in Manhattan but railroad stations as well. Cassatt's proposal for a through freight route contemplated the use of car floats across New York Bay, between Greenville and Bay Ridge. From Bay Ridge, freight trains would run over Long Island rails to Island Pond Junction, where a transfer would be made to the Connecting Railroad's tracks for the remainder of the journey to Port Morris. Currently, the Pennsylvania ferried its freight cars the entire distance between Port Morris and the north Jersey docks, a most laborious procedure.[66]

The idea of a connecting railroad had been around for some time. Cassatt himself had embodied it in his 1891 report to President Roberts. Shortly after Lindenthal resigned from the board of engineers, Cassatt put him to work drawing blueprints for a bridge across Long Island Sound at Hell Gate, not far from Port Morris. And surely a connecting railroad was on the PRR president's mind when he spoke with stockholders about that "convenient connection" with the New Haven, for which the acquisition of the Long Island was to pave the way.

Cassatt's goals could not be fully achieved without relying on the use of electric motive power. Steam locomotives could never be used to haul passenger trains through the East River tunnels to Penn Station. The New York Connecting Railroad had existed on paper since 1892. It had not gotten beyond that stage primarily because it could find no alternative to the steam engine.[67] Freight trains could perhaps lumber through Long Island behind steam locomotives to and from Bay Ridge, but even there public pressure was mounting in opposition to excessive smoke and dirt. Now that both the Pennsylvania and the New Haven had firmly committed themselves to electrification, a connection establishing through freight and passenger service by way of Long Island seemed a certainty. Establishment of such a link to the north and east would be a most fitting climax to the PRR's long-contemplated New York improvement program.

2

....

The Search for Motive Power

To Cassatt and other top officials of the Pennsylvania, electrification seemed to fit the needs of the railroad's New York extension in almost every way. It offered numerous advantages over steam locomotion: speed, power, ease of maintenance, and most importantly, lack of combustion exhaust. Only two drawbacks attached themselves to the use of electric traction. One of these was the high cost of installing the system. Electric operation required the construction of expensive power generating and distributing gear. On the other hand, the money thus spent was dwarfed by the millions of dollars the railroad was lavishing on tunnel and terminal improvements. Moreover, the high first cost would be offset to some extent by the reduced operating expenses, if the electrification were successful.

That "if" related to the second of electric traction's deficiencies. In 1901, there existed only three practical examples of steam railway electrification which might serve as guides for the PRR: England's City and South London system, the Orleans extension in Paris, and the B&O's Howard Street tunnel in this country. None of these operations compared in size or capacity to what the Pennsylvania had in mind. Therefore, the railroad would have to custom build its New York extension electrification without having the benefit of other lines' experience.[1] The electrification of the New York Central and the New Haven would provide a few guidelines, to be sure; but these installations were not completed until the PRR's own traction system was in the final stages of development.

A railway electrification based so heavily on technological innovation was likely to suffer from a multitude of deficiencies at least over the short run. Under the very best of circumstances, the New York extension electrification was doomed to experience a lot of minor "teething problems," once it was made operational. At worst, enough technical setbacks could occur to put the Pennsylvania in a very awkward and embarrassing situation.

The key to avoiding such a predicament was to have the railroad conduct its research and design studies in a very deliberate manner. Such a methodical approach was nothing new to the PRR. Its mechanical engineering staff at Altoona had been successfully custom designing the road's steam locomotives for several decades. On those occasions when the Pennsylvania did contract with an outside builder, it was usually for the purpose of mass producing engines from blueprints that the railroad's own engineering people had drawn up.[2]

Now, however, the Pennsylvania needed electric motive power, not steam. Its mechanical department naturally lacked familiarity with electric traction. This explained the addition of George Gibbs to the railroad's board of engineers. Gibbs was born in 1861 in Chicago. After receiving a degree in mechanical engineering from Hoboken's Stevens Institute of Technology, he worked for several years as an assistant to Thomas Edison. In 1895, when the Westinghouse Electric and Manufacturing Company and the Baldwin Locomotive Works decided to pool their resources in the manufacture of electric locomotives, the two companies retained Gibbs as chief engineer of the European affiliates of the Westinghouse firm. Upon assuming his new duties on the PRR's engineering board, Gibbs found his task to be more complicated than either he or Cassatt had anticipated. In 1903, the New York Central announced that it would electrify its New York terminal operations using the trustworthy direct current system. Not long thereafter, the New Haven disclosed its intention to electrify with single-phase alternating current. The New Haven's decision stunned the railway industry. A few European railroads were just beginning to experiment with poly-phase a.c. systems. No railroad anywhere had as yet tested a single-phase system. Even the New Haven's own electrified branches employed direct current.[3]

The Pennsylvania Railroad was not in the habit of casting aside proven, workable technology. Still, it was not inclined to overlook any technological innovations that could yield potential benefit. Consequently, the railroad decided to explore the merits of both a.c. and d.c. before committing itself to one or the other. The New York Central had created a special electric traction commission to determine the general features of its electrification. This panel, chaired by Central Vice-President William J. Wilgus, included in its membership such well-known consulting electrical engineers as Frank Sprague, Bion J. Arnold, and the PRR's own advisor, George Gibbs. The Pennsylvania, clinging to its tradition of handling its own engineering work whenever possible, did not bring in any consultants other than Gibbs. President Cassatt appointed him chairman of an electric locomotive committee, the other members of which were PRR mechanical engineer Axel S. Vogt, and Alfred W. Gibbs (a cousin of

George) and David S. Crawford, superintendents of motive power, Lines East and West, respectively. Cassatt ordered the group to conduct a thorough investigation of the a.c. and d.c. systems and then to select the one most appropriate for the PRR's needs.[4]

The direct current method of power generation and distribution had been developed first. Its equipment was simple, and its maintenance costs were low. Simplicity of construction contributed to the long life and dependable service of d.c. motors and generators. In these early years, direct current flowed at low voltage levels, minimizing the danger of injury to those persons unaware of the potency of electricity. This last factor alone was enough to convince Edison of the superiority of direct current. His Edison General Electric Company marketed a wide variety of d.c. equipment. In 1892, the Edison firm and the Thomson-Houston Electric Company—the only other major manufacturer of d.c. equipment—merged to form the General Electric Company.

In the meantime, rapid progress was being made in the perfection of the alternating current system. In 1886, George Westinghouse, already having made a fortune in the railway air brake and signal businesses, established the Westinghouse Electric and Manufacturing Company to investigate and exploit the commercial possibilities of alternating current in the United States. One of the company's first acts was to purchase European patent rights on some crude a.c. equipment. When the English-designed transformer proved defective, Westinghouse and his engineers soon developed one of their own. In the fall of 1886, Westinghouse Electric installed the world's first commercially successful system of a.c. generation and transmission at Buffalo, New York.[5] That event triggered a heated battle between Westinghouse and its Edison and Thomson-Houston rivals over which system offered the greater benefits.

Direct current could not be transmitted over long distances without suffering a severe drop in voltage. The use of complicated and costly substations ameliorated this condition somewhat but did not eliminate it entirely. In addition, the use of such low voltage levels meant that a high amperage (or a large amount of current, as distinguished from the "pressure" of the current, to which voltage refers) was needed. To provide an adequate conductor for this current, a great deal of copper was required. The characteristics of alternating current, including its ability to flow at much higher line voltages than direct current, allowed it to be transmitted over far greater distances without the supportive substations. Not only did the higher voltages eliminate the drop-off phenomenon, they also reduced the demand for copper. As the voltage level increased, the amount of conductive wire required to carry it decreased. The heart of the a.c. system was the transformer. It increased (stepped up) the voltage at the generating plant so that the current could be transmitted over many miles economically. Another transformer reduced (stepped down) the line voltage before the current was fed to lamps and motors. With a growing number of customers located increasingly farther away from central generating stations, utilities soon récognized the superiority of alternating current for most residential and commercial uses. General Electric tacitly admitted defeat in 1896 when it concluded a patent agreement with Westinghouse. That arrangement gave each company the freedom to manufacture electrical equipment using the other's patents. Both firms thereafter produced a wide assortment of a.c. and d.c. motors, generators, and related gear.[6]

In only one area did this "battle of the systems" sputter on. Many electrical engineers refused to concede that alternating current held the advantage in railway applications. Over this point, Westinghouse and GE continued to clash bitterly. Even though each firm had the opportunity to market either system, GE clung tenaciously to direct current while Westinghouse just as stubbornly held out in favor of alternating current.[7]

The d.c. series motor was unquestionably the most efficient for railroad use. In this type of motor (so named because its armature and field were wound in series), the field varied automatically with the load. As the load increased, the field strength increased. This led to a decrease in speed but an increase in tractive effort. Conversely, a decrease in load brought a decrease in tractive effort but an increase in speed. The series motor, therefore, adjusted itself readily to the varying tractive effort required in railway work. It permitted the same locomotive to haul either a heavy train on level track, or a train of normal weight up steep grades and around sharp curves. The relationship between field and load possessed much the same effect as shifting gears did in an automobile, that is, an unusual amount of effort could be exerted by a relatively small motor. Complimenting the simplicity of the series motor's design was its ability to operate at full-line voltage (generally about 600 volts) without having to resort to any transformer. Street railway and interurban equipment had been using the d.c. series motor for many years. By the early 1900s, railway men were well versed in its operation and repair.

Unfortunately, the characteristics of a d.c. motor of any kind confined its operation to low voltage levels. As railroads began to run longer and heavier trains, they required more current than an overhead wire type of distribution system could deliver. Railroads then had to turn to one of several methods of third-rail distribution.[8] Edison's supposedly "safe" direct current became lethal, exposed as it was at ground level, despite attempts to shield the energized rail from the unwary. But the inadequacy of direct current for long-distance transmission was by far its greatest handicap. This was not a matter of much regard for installations spanning only a few miles. Any railroad contemplating the use of direct current for heavy-duty electrification covering more than just a few short route-miles, however, would have to be prepared to construct an elaborate series of substations. Even multiple generating stations might be necessary, depending upon the distance.[9]

Alternating current was not without its share of disadvantages. The a.c. series-commutator motor, although similar in design to the d.c. series variety, was very bulky. Transformers and related control equipment occupied an inordinate amount of space and were very heavy. Owing to commutator limitations, the series-commutator motor could not produce as high a starting tractive effort (the effort required to move a standing train). Furthermore, maintenance costs were significantly higher for the more complex a.c. systems, which also carried a higher first cost. William Bancroft Potter, head of GE's railway division, contended that alternating current was "applicable only to exceptional conditions." Railroads that operated heavily laden trains over severe mountain grades, he conceded, could use alternating current to good effect because of the a.c. motor's ability to produce more horsepower. The greater weight of a.c. locomotives would also be of some benefit. Under most other circumstances, direct current was far superior.[10]

The New York Central, with its great diversity of short-haul traffic to and from Grand Central Station, agreed with Potter. In 1903, it awarded its electrification contract to GE with hardly any consideration of plans for an a.c. system submitted by Westinghouse. The road's electric traction commission regarded the use of alternating current as impracticable at this stage of development. General Electric's Asa F. Batchelder designed a 2,000-horsepower prototype locomotive. In 1905, the Central subjected this engine to an extensive series of trials on a short stretch of the road's main line located within sight of General Electric's Schenectady headquarters. Never before had electric motive power been tested so rigorously. The railroad expressed a particular interest in comparing the electric engine's power and acceleration qualities against those of one of its finest steam locomotive types.[11] At year's end, the Central, completely satisfied with the performance of its new locomotive, placed an order with GE for 35 additional units. The railroad designated these as Class T. On September 30, 1906, T-motor No. 3405 pulled the first locomotive-drawn electric train into Grand Central Station. So pleased was the Central with these engines that it added a dozen more (their model now changed to Class S) to its roster in 1908. It assigned most of these locomotives to long-distance passenger trains. A fleet of 180 multiple-unit cars handled most of the local transit.[12]

The New York Central chose the proven technology of direct current and was rewarded handsomely. Maintenance costs for its electric rolling stock fell far below that of even its most modern steam locomotives. Service increased in dependability and speed. The people of Manhattan and the Bronx luxuriated in fresh, clean air for the first time in memory. Only the faintest hint of a dark cloud appeared on the Central's electrification horizon. Significant expansion of the 650-volt third-rail system would not be feasible.[13] If the railroad ever hoped to have electrically powered trains running all the way to Albany, for instance, it would be forced to make substantial—and costly—modifications to the present system.

At first glance, logic would seem to have dictated that the New Haven follow the New York Central's lead and adopt an identical d.c. electrification. The New Haven already had experience with direct current on branch lines in Connecticut and Massachusetts. More than a few eyebrows were raised in traction circles, then, when President Charles S. Mellen of the New Haven announced in 1905 that his railroad had accepted a proposal from Westinghouse for a single-phase, 11,000-volt a.c. system. The first stage of the electrification would encompass the 21 miles between Woodlawn Junction and Stamford, Connecticut, the road's most heavily trafficked commuter zone. Whereas the New York Central had given little thought to the idea of long-distance electrification, the management of the New Haven held that consideration uppermost. As New Haven Vice-President E. M. McHenry remarked, the decision to opt for alternating current "was based upon a study of the subject as a whole rather than upon the solution of the terminal problem only." If the a.c. system were successful, the railroad expected to extend it eastward in the near future to the city of New Haven, 75 miles from Woodlawn Junction, and eventually to Boston, 232 miles beyond New Haven.[14]

The major problem confronting Westinghouse and New Haven engineers in 1905 was the sheer novelty of alternating current in railway applications: only one railroad (actually, an interurban) had as yet dared to electrify with it. Westinghouse had supplied single-phase equipment for the Indianapolis and Cincinnati Traction Company's 3,000-volt 40-mile line between Indianapolis and Rushville, Indiana. Operations there were not scheduled to begin until sometime in 1906. So new was alternating current in railway use that Westinghouse's East Pittsburgh works had just outshopped the very first single-phase a.c. locomotive built in America, a 1,300-horsepower twin-unit box cab. This engine was to be used for gathering data for subsequent construction of New Haven (and possibly PRR) a.c. locomotives.[15]

The New Haven took delivery of the first of 35 Class EP-1 locomotives from Westinghouse late in 1905. These units, patterned after the original Westinghouse prototype, weighed 102 tons and sported a B-B wheel arrangement. Each was capable of generating 1,420 continuous horsepower. Like the Central's S-class engines, the EP-1's were designed to wheel 250-ton express trains at speeds up to 65 miles per hour. Their series-commutator motors performed well on either type of current. On each locomotive was mounted a pair of third-rail shoes that could be lowered to collect direct current between Woodlawn and Grand Central.[16]

Westinghouse and New Haven electrical engineers braced themselves for a flurry of malfunctions during the first few months that the a.c. system was in operation. Their expectations did not go unfulfilled. The EP-1's exhibited a tendency to oscillate from side to side at high speeds, playing havoc with the track and inviting a serious derailment. The overhead contact wire with its two supporting cables proved to be too rigid.[17] This lack of flexibility resulted in the wire being kinked and even broken by the locomotives' pantographs, thus interfering with current collection.[18] The a.c. generators at the New Haven's Cos Cob (Connecticut) powerhouse could develop no more than two-thirds of their normal, continuous horsepower before they overheated. The exhaust blasts of steam locomotives kept corroding the insulators on the 11,000-volt wiring. These and numerous lesser difficulties had so crippled the New Haven's main line that by the spring of 1908 the railroad was seriously considering total cessation of electric operation until the problems could be resolved.[19]

The Pennsylvania's electric locomotive committee began its deliberations at about the same time the New Haven was announcing that it had selected alternating current for its electrification. Since the New Haven's misfortunes still lay in the future, Gibbs and his committee were prepared to give equal consideration to either system, just as Cassatt had instructed.[20] Even if the Gibbs group had been able to foresee the ordeal of the New Haven, later events would show that it would not have been deterred in its study of the feasibility of alternating current for the New York extension. Indeed, why not let the New Haven incur the expense and hardship of perfecting the a.c. system, and have the PRR reap any technological harvest that resulted? For that matter, the Pennsylvania had already forfeited the honor of pioneering an extensive d.c. system to the New York Central. Here again, however, the PRR could only gain by the other railroad's experience. The Pennsylvania was fortunate in that it did not expect to have its extension operational until 1910 or 1911.

The New Haven and the New York Central by law had to convert to electric motive power before July 1, 1908.

On the other hand, conditions facing the Long Island Rail Road necessitated that line's adoption of electric traction as soon as possible. The Pennsylvania decreed in 1904 that the Long Island Rail Road's electrification should utilize 650-volt direct current. There was little time to experiment with an a.c. system. Nor was there any pressing need for the use of alternating current. Geographic limits were such that the Long Island would never become a long-haul carrier. As soon as the East River tunnels were completed, the Long Island would extend its third rail all the way to Penn Station.[21] This extension argued more strongly for the Pennsylvania's own use of direct current than any other factor, with the exception of the proven reliability of d.c. equipment. By adopting direct current, the PRR could avoid duplication of electrical gear while simultaneously effecting a much smoother interchange of rolling stock with its affiliate.

No matter which system the Pennsylvania chose, there could be little doubt that it would ultimately award the general electrification contract to Westinghouse. In the event the railroad's tests led it to select alternating current, it would have little choice but to turn to the champion of alternating current, just as the New Haven did, to supply the equipment. Westinghouse was also knowledgeable and experienced in the realm of direct current, having designed and manufactured the equipment for the Long Island's electrification. A personal familiarity between Westinghouse and PRR personnel existed as well. Engineers of both companies cooperated in electrifying the Long Island. And George Gibbs was simultaneously chief electrical consultant for the Long Island Rail Road, head of the Pennsylvania's traction program, and first vice-president of Westinghouse, Church, Kerr, and Company, the consulting engineering arm of the Westinghouse Electric and Manufacturing Company.[22] The PRR could not afford to overlook, either, the fact that Westinghouse, with its sprawling East Pittsburgh works, was an important on-line customer.

In spite of its seemingly favorable relationship with the Pennsylvania Railroad, Westinghouse took no chances that it still might lose the forthcoming contract. In 1905, it reached an agreement with the railroad whereby it provided the electrical gear for two experimental d.c. locomotives, the bodies of which the PRR built at its own Juniata (Altoona) shops. The remarkable aspect of this accord was the willingness of Westinghouse to furnish all equipment free of charge. In addition, it put its railway engineering staff at the railroad's disposal, similarly free of any cost to the PRR.[23]

The Pennsylvania completed construction of the two experimentals late in 1905. The first unit, to which the railroad assigned No. 10,001, was an 88-ton box cab mounted on a pair of articulated trucks, giving it a B+B wheel classification. Its four axle-hung (also known as nose-suspended) motors, geared directly to the axles, each developed 350 horsepower. The second engine, No. 10,002, weighed 98 tons but rode on the same kind of B+B articulated trucks as No. 10,001. A pair of 320-horsepower motors propelled one truck, while two 300-horsepower motors powered the other. All four traction motors employed a gearless quill drive rather than the geared drive common to nose-suspended motors. The quill was a hollow shaft having a diameter about 2 inches greater than the axle. It fit over—but did not touch—the axle, much in the manner of a sleeve. At one end of the quill was a circular metal plate, or "spider," from which projected a number of pins. The pins were so situated that they projected outward between the spokes of the driving wheels. As the quill (which carried the motor armature) rotated, it turned the wheel and axle. The PRR did not radically depart from past practice by placing a gearless quill drive in No. 10,002. The Baltimore and Ohio's electric locomotives of 1895 and the New Haven's EP-1's utilized an identical concept. Geared drives had found widespread use in interurban cars but had proved unsatisfactory for high-speed service. The Pennsylvania was primarily interested in determining whether the gearless type would be more suitable.[24]

Since the Long Island's electrification had become fully operational by this time, the PRR sent its two experimentals there for more thorough testing than either the East Pittsburgh or Altoona facility allowed. The railroad had designed each locomotive to run at a maximum speed of 45 miles per hour, for it still contemplated ending the electrification at Bergen Hill rather than across the Meadows at Harrison. Higher speeds, therefore, were not necessary. The trials had barely gotten under way, however, when the Pennsylvania negotiated its second pact with the Hudson and Manhattan and moved the western terminus of the electrification to Harrison. The railroad then modified Nos. 10,001 and 10,002 to permit speeds up to 65 miles per hour. In subsequent testing, the two units developed a rhythmic sideway in the higher speed ranges. The phenomenon resembled that just discovered by New Haven engineers while they tested that road's EP-1's. The severe lateral pressures exerted by the locomotives wore out rails at an alarming rate.[25]

The PRR could never tolerate such performance if its electrification were to be successful. At a time when the Pennsylvania and the New York Central were locked in fierce competition for the New York-Chicago passenger market, a few minutes counted for thousands of dollars. To spend millions of dollars on terminal improve-

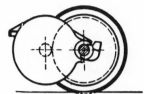

1—Axle Hung, Nose Suspended
Motors. Single Reduction Gear.

2—Gearless.
Armature on Axle.

3—Gearless.
Armature on Quill.

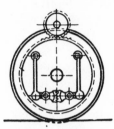

4—Geared Flexible.
Motors on Frame.

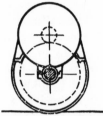

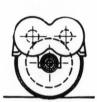

5, 6—Geared Quill, Motors and Quill on Frame.

COLLECTIVE DRIVES

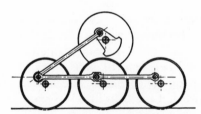

7—Direct Side Rod Drive

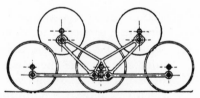

8—Direct Scotch Yoke Drive

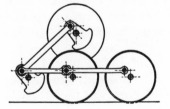

9—Jack Shaft Side Rod Drive

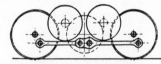

10a—Geared Jack Shaft Side Rod Drive

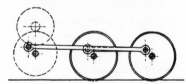

10b—Geared Jack Shaft Side Rod Drive

Types of drives used by electric locomotives. *(Locomotive Cyclopedia)*

ments only to lose the speed race to the Central was unthinkable. The PRR simply had to have high-speed electric locomotives.

Alfred Gibbs suggested that the problem of lateral stresses could best be remedied if the effects of steam and electric locomotives could be studied side by side. "The greatest difficulties of the electric locomotive, other than financial, are mechanical rather than electrical," he declared. Therefore, would it not be wise to pit the mechanically flawed electrics against proven steam types, with the idea of transferring certain qualities of the steam types to the electrics?[26] His recommendation met with favorable reaction. The only impediment to its implementation was the fact that the in-depth testing Gibbs had in mind could never be carried out on the heavily trafficked Long Island. The railroad put its two electric engines in storage until a more convenient site could be found.

By early 1907, George Gibbs and his electric locomotive committee decided to resume the tests on a 7-mile segment of the West Jersey and Seashore Railroad, near Franklinville, New Jersey. The WJ&S, a wholly owned subsidiary of the Pennsylvania, ran between Camden and Atlantic City. In 1906, the PRR had contracted with General Electric to install a 650-volt d.c. third-rail system on one of the two WJ&S routes linking the two cities in order to expedite traffic to and from the shore resorts.[27] During the summer of 1907, PRR crews modified a short stretch of the southbound track between Camden and Franklinville in anticipation of the upcoming tests. Eighty cast-steel ties were placed over a distance of 165 feet of track. On top of each of the special ties and flush with the outer face of one of the rails was fastened a hardened steel ball, about 1 inch in diameter. Running parallel to the rail and clamped between the balls and the outer end of each tie was a long strip of boiler plate. The lateral force of a train passing over the rails pressed the balls into the plate, which was moved longitudinally after each passage in order to record a new impression. After each ball made a series of impressions (usually about 30) in the plate, workmen used a mechanical testing device to calibrate the depth of each. An average was then taken which accurately indicated the lateral impact made by any given locomotive.[28]

The PRR brought down Nos. 10,001 and 10,002 from Long Island early in September. These engines then proceeded to run but a few desultory trials, more to sharpen the accuracy of the measuring instruments than to record any rail wear.[29] The railroad delayed the commencement of all-inclusive testing until it could ready an a.c. locomotive to match against the original pair of d.c. experimentals. With as little as three years left before the opening of the New York extension, the electric locomotive committee felt that experimentation with alternating current should not be delayed any longer. Since a third engine had to be built, the committee chose to equip it with a wheel arrangement different from that found on the first two units. In this way, the railroad could weigh the merits of both systems and at the same time try to resolve the lateral thrust problem.[30]

In order to hasten construction, the PRR commissioned the Baldwin Locomotive Works of Philadelphia to build the carbody from blueprints prepared by George Gibbs and his colleagues. Baldwin shipped the empty shell to East Pittsburgh, where Westinghouse outfitted it with the proper electrical gear. What emerged from the Westinghouse plant in July 1907 was PRR electric locomotive No. 10,003. Its two traction motors, coupled to the driving wheels through gearless quill drives, generated a total of 750 horsepower. In place of the B+B running gear was a novel 2-B arrangement, that is, two guiding axles followed by two powered axles. This compared to a 4-4-0 steam locomotive. The locomotive committee was gambling that since steam locomotives having this wheel configuration (Class D on the PRR) showed no extraordinary rail wear, an electric locomotive having an identical pattern would perform just as satisfactorily. Alfred Gibbs was especially confident on this point. He argued that the sideway of the first two electrics, as well as that of the EP-1, resulted in part from these units having a symmetrical wheel arrangement. Experience with steam locomotives, most notably 2-4-2 types, had demonstrated that a combination of short wheelbases, symmetrical wheel arrangements, and heavy overhanging weights distributed through the length of the engines invariably caused undue lateral pressures. This lesson seemed to have been forgotten when designing electric locomotives, which had heretofore evolved from streetcars rather than from steam locomotives.[31]

Before sending No. 10,003 to Franklinville, the PRR and Westinghouse jointly put the black monster on display for newsmen at East Pittsburgh. The reporters could hardly help but be impressed as the locomotive accelerated from stop to 65 miles per hour and back to stop again, all in the space of 1¼ miles. They were equally awed by the eerie silence with which the engine went about its work. The gleaming No. 10,003 was an iron horse different from any the general public (and most railroad men, for that matter) had ever seen.[32] With the curiosity of the press gratified, the railroad ordered Westinghouse to conduct a few more minor tests with the locomotive. In late October, the electrical manufacturer turned the machine over to the PRR, which then shipped it without further ado to Franklinville.

There No. 10,003 joined an imposing array of steam and electric motive power. Supplementing the Pennsylvania's own traction roster of three units was an EP-1, which the railroad had borrowed from the New Haven. Representatives from two of the PRR's finest steam classes, D16b and E2, were on hand, too. Since the a.c.-powered No. 10,003 could not ordinarily operate over the direct current system of the West Jersey and Seashore, the PRR installed a motor-generator (m-g) set in an express car, fitted the car with third-rail shoes, and coupled it immediately behind the new locomotive. Direct current collected from the third rail drove a d.c. motor, which in turn powered an a.c. generator, thereby furnishing alternating current for No. 10,003's traction motors.[33]

For two weeks in mid-November, the tranquillity of the south Jersey countryside was shattered as the Pennsylvania Railroad loosed its herd of mechanical beasts over the path of the WJ&S. The various engines ticked off mile after mile, often at speeds bordering on 100 miles per hour for steamers and in excess of 80 miles per hour for each of the electrics. On each of the two Saturdays, the 5 miles between Franklinville and Clayton were enclosed by a solid wall of expectant faces and waving arms, as spectators from as far away as Philadelphia and New York gathered to witness the proceedings. Many of the onlookers returned home with the mistaken notion that the PRR was hurling electricity against steam in a titanic duel, the outcome of which would settle the fate of motive power on all America's great railroads. If the Pennsylvania's true aim was not so grandiose, it was, nonetheless, sufficiently important to justify spending $250,000 on the Franklinville tests—by far the largest sum expended to that date on any kind of railway experiment.[34]

The results of the November spectacular were startling. Both 4–4–0 and 4–4–2 steam types registered an average impression in the boiler plate of 0.0123 inch. The railroad used this as the standard by which to gauge the performance of the electrical units. Locomotive No. 10,001 exhibited an average indentation of 0.0216 inch, and No. 10,002 averaged 0.0291 inch. The New Haven's EP-1 made an average impression of 0.0149 inch. This last reading "could not be considered bad," according to Alfred Gibbs, but it was still greater than the Pennsylvania preferred. It was also the result of incomplete testing because the PRR soon had to return the locomotive to its owners. Locomotive No. 10,003 indented the plate an average of 0.0151 inch. If the five runs made in reverse with the idler truck free were not included, however, the average dropped to 0.0123 inch, the exact impression recorded by the steam engines.[35]

These figures convinced the electric locomotive committee of the value of a nonsymmetrical wheel arrangement. The a.c. versus d.c. dilemma, on the other hand, remained to be resolved. At Franklinville, the railroad evaluated the tracking qualities of its locomotives. It made no attempt to gather data on the performance of the locomotives' electrical systems. The doubts about a.c. traction centered more on such external features as collection devices and catenary construction than on the internal functioning of the locomotive it-self. The catenary, an overhead system for distributing current, con-sisted of one or more contact wires, as well as supporting wires, braces, poles, and related gear. The contact wire assumed a constant height above the rails and was suspended from one or more nonenergized messenger wires, which were attached to a series of rigid cross braces, which in turn were attached to poles and ran per-pendicular to the rails. When suspended under uniform load, a mes-senger wire took the shape of a true catenary curve—hence the name for the system as a whole.

Could alternating current electrification as a whole withstand the punishment of heavy traffic over a long period of time? To answer this question, the Pennsylvania had to return to the Long Island Rail Road. During the summer of 1908, crews erected an 11,000-volt, 25-cycle, single-phase overhead wire along a 5-mile stretch of freight-only trackage between Garden City and Babylon. Various types of catenary were used. Included on the line was a 950-foot "dummy" tunnel which reproduced the clearances of the Hudson River tunnels. Over these 5 miles ran No. 10,003 to and fro, night and day, week after week. Often trailing behind were two work cars on which pantographs had been positioned to facilitate the study of several methods of current collection. The Long Island supplied a number of old coaches so that realistic train weights could be approximated.[36]

The results of these trials must have disappointed a.c. proponents, but they came as no surprise to George Gibbs. While No. 10,003 acquitted itself decently, the distribution and collection equipment could not endure such vigorous use. The tests, said Gibbs, "demon-strated the need for considerable further experimental work to adapt the a.c. system to tunnel and yard conditions."[37] On December 1, 1908, the railroad discontinued all testing. A few days later, it an-nounced that it was going to electrify its New York extension with direct current.[38]

In practice, the PRR would utilize a combination a.c. / d.c. system similar to those already used by the New York Central and Long Is-land railroads. The Long Island Rail Road generated 11,000-volt, 25-cycle, single-phase current at its Long Island City powerhouse. From there, it distributed the current through an overhead wire network to four strategically located substations. At each substation, motor-generator sets converted alternating current to 650-volt direct current, which was then fed into the third-rail system. Such an ar-rangement capitalized on the transmission efficiencies of alternating current and the operational simplicity of d.c. traction motors. A compromise between the two systems did not by any means eliminate all the inherent faults of each, however. The Long Island had to build several portable substations (motor-generator sets mounted on railway cars) for use on the most heavily traveled routes. When traffic reached peak levels, as during rush hours, voltage tended to fall unless the railroad rushed in with a couple of extra substations to provide more current. Once the Pennsylvania had committed itself to direct current, it decided to enlarge the Long Island City generating plant and plug its own substations directly into the circuits of its subsidiary. George Gibbs estimated that three more Westinghouse-Parsons 5,500-kilowatt turbogenerators would be needed at Long Island City and several additional substations would be required to cover the PRR's electrified zone.[39]

Ironically, the Pennsylvania disclosed its intention to use direct current just as W. S. Murray, chief of electric traction for the New Haven, was reporting to a meeting of the American Institute of Electrical Engineers that "history sustains the undeniable truth that alternating current is the preferred agent for the transfer of electricity where either distance or capacity is involved. A railroad involves both."[40] The New Haven had found a cure for nearly all its a.c.-related problems by this time. Emulating PRR practice, for example, the road had added a guiding axle to each end of the EP-1's, giving them a 1-B-B-1 wheel configuration and bringing an end to their tracking difficulties.

A few years later, when discussing railroad electrification with Murray, Alfred Gibbs acknowledged that even as the PRR was adopting direct current, it still recognized alternating current to be the superior system. The Pennsylvania simply could not afford to find itself in a position, as the New Haven almost did, of de-energizing its electrification until it had gotten rid of all the imperfections. "Any serious operating failures," said Gibbs, "would have jeopardized the whole investment and put electrification back for years."[41]

Between the time the Franklinville tests had been concluded and the Long Island tests begun, the two Gibbses, Vogt, and Crawford had sketched plans for a new locomotive employing the 2-B arrangement of No. 10,003. One of these units, whether alternating or direct current, would not be sufficiently powerful to pull a train at the speeds the PRR demanded. To surmount this handicap, the designers proposed placing two units back to back to form a single hinged or articulated locomotive having a 2–B+B–2 classification. Initially, the locomotive committee considered using a nonarticulated frame. The engineers suspected articulation as being the cause of a "snaking" motion which compounded the problem of rail wear. The committee only reluctantly decided in favor of articulation, reasoning that the long wheelbase (over 60 feet) would be too inflexible if not hinged.[42]

The PRR's mechanical engineers (and those of other railroads) had discovered many years before that incorporating a high center of gravity into a steam locomotive improved its ride and reduced the impact of side blows to the rails. The high center of gravity acted as an inverted pendulum to counterbalance the sidesway produced at the axle level of all locomotives. Both the nose-suspended and the gearless quill drives fit between the driving wheels, automatically giving the electric locomotive a low center of gravity. PRR engineers believed that this arrangement intensified the lateral pressures which their experimentals exerted. When drafting blueprints for the new 2–B+B–2, then, they raised the traction motors to the level of the main deck of the locomotive. Relocating the motors conferred two advantages on the new machine. A higher center of gravity meant increased tracking stability, of course. Having the motors mounted on the deck of the engine also meant that the railroad could install larger, more powerful motors. Formerly, traction motors had to be small enough to fit into the confined space between the driving wheels, a stipulation that naturally put a limit on horsepower. Now the PRR could utilize the full width of the locomotive cab.[43]

Situating the motors on the deck or frame of the locomotive posed the problem of how to connect the armature to the drivers. A gearless quill mechanism was impractical now that the motor was so far above the axle. The locomotive committee discarded the possibility of gearing as being too intricate and too costly. Instead it settled on an arrangement of jackshafts and side rods. To each end of the armature was attached a crank. A pair of rods connected the cranks to a common jackshaft. Another set of rods coupled the jackshaft to the driving wheels in imitation of the main rods on a steam locomotive. This system possessed an added advantage in that one motor could drive more than one axle, since all the main axles were connected by rods. Side-rod drive eliminated the slipping of driving wheels which sometimes occurred because of unequal distribution of weight in locomotives (such as Nos. 10,001 and 10,002) which had independently powered axles.[44]

Plans for this fourth experimental engine were completed well before No. 10,003 had wound up its trials on the Long Island. In early November 1908, the Pennsylvania awarded a $5 million contract to Westinghouse for an unspecified number of electric locomotives of the 2–B+B–2 type. Whether the machines were to be a.c. or d.c. depended on the a.c. endurance tests then under way, but the locomotive could be adapted easily for either system. In December, with the PRR now definitely prepared to adopt direct current, railroad and Westinghouse engineers met to draft final specifications for the locomotives' electrical gear and to determine just how many units would be needed. In May of the following year, the Pennsylvania ordered 24 two-unit sets. Westinghouse was to manufacture the electrical components. The railroad's Juniata shops would provide the carbodies and mechanical parts. The PRR rushed the first set to completion so that it could be thoroughly tested. In this way, any modifications to the original design could be embodied into the remaining 23 locomotives while they still sat on the erecting floor.[45] Five months later, the Pennsylvania took delivery of the first locomotive, which was dispatched immediately to Long Island to undergo the customary scrutiny by the engineering staff. Throughout the winter of 1909–10, Westinghouse and railroad engineers monitored its performance. They focused particular attention on the problem of rail wear. Despite many improvements in their construction, steam locomotives still punished the rails because of the unbalanced effort exerted by the driving wheels. This imbalance stemmed from the unevenness with which the piston of the reciprocating engine applied the power stroke. Railroads had little choice but to tolerate this problem. Indeed, the PRR was delighted when at Franklinville it reduced the lateral impact of its electrics to something comparable to that produced by its steamers. The question before the road's mechanical engineers now was, would this newest of experimental electrics with its side-rod drive pound the rails in the same manner as the steam locomotives from which it was derived?

To the great relief of all concerned, the application of jackshafts and side rods actually reduced rail wear. The drawbar pull of the new locomotive was constant and uniform. The electric motor, unlike the steam engine, produced an unvaryingly even motion. The uniform effort at the armature translated into uniformity at the driving wheels, just as the unbalanced effort at the piston of a steam engine translated into imbalance at the drivers. The Pennsylvania, satisfied that no significant changes need be made to the prototype, signaled Westinghouse to finish installing the internal gear on the other 23 locomotives.[46]

Steamers having a 4–4–0 wheel arrangement the PRR denominated Class D. Since the new electrics were essentially two 4–4–0's, the railroad assigned them Class DD1. Each unit carried a 1,580-horsepower Westinghouse Type 315-A series-wound motor. A two-unit set thus packed a total of 3,160 continuous horsepower, making it the world's most powerful locomotive, steam or electric.[47]

As the DD1's began rolling east during the summer of 1910, the Pennsylvania put the finishing touches to Penn Station and adjoining tunnels. Unfortunately, A. J. Cassatt did not live to see his dream become reality. He died on December 12, 1906. On August 1, 1910, his successor, James McCrea, and various PRR officials and members of the Cassatt family unveiled a statue of the deceased president at

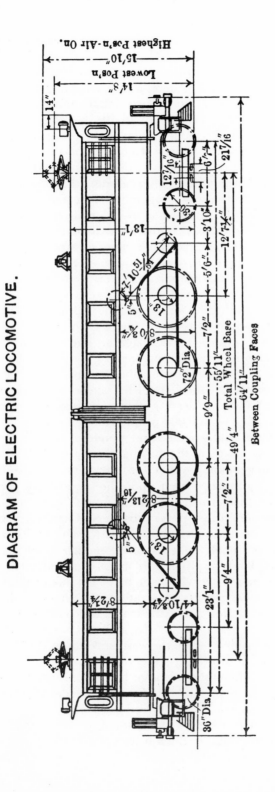

DIAGRAM OF ELECTRIC LOCOMOTIVE.

Line drawing of a two-unit DD1. (American Society of Civil Engineers)

Pennsylvania Station and formally declared the structure open to the public.[48]

Trains did not begin hauling passengers to and from the terminal until September 8. On that date, a brace of Long Island m.u. cars inbound from Jamaica coasted to a halt alongside a platform jammed with civic dignitaries, railroad officials, and the general public to become the first regularly scheduled passenger train to arrive at Penn Station. For the next two months, the PRR permitted only Long Island trains to use the new station. This gave the Pennsylvania time to care for last-minute details regarding the handling of its own trains to and from the west. Beginning November 13, the railroad staged a two-week dress rehearsal in order to provide its personnel with at least a modicum of on-the-job experience prior to greeting the traveling public. Eastbound steam-powered trains deposited their passengers and exchanged motive power at Harrison Transfer and then ran empty (except for train crews) into Manhattan. There terminal workers went about their normal duties just as if each train had brought in a full complement of riders. Next the DD1's pulled the trains out to Sunnyside yard, where each car was cleaned and serviced before being sent west again. The railroad dealt with westbound trains in a similar manner, with patrons boarding the empty cars at Harrison Transfer and Newark. At 12:02 A.M., on Sunday, November 27, the first PRR train with paying passengers aboard, a local bound for Perth Amboy, departed from Pennsylvania Station. Later that day, 100,000 people visited the terminal to marvel at the mechanical and architectural wonders wrought by the railroad.[49] Nearly a decade in the making, the New York extension had finally begun to generate revenue for its owners.

The PRR congratulated itself upon the successful grand opening of what its publicists termed "a permanent monument to the mastery of science over the greatest physical barriers of nature."[50] At a cost of over $160 million, the railroad had at last anchored its eastern terminus in the profitable bedrock of downtown Manhattan. It had surmounted all kinds of physical, financial, political, and technological obstacles. The greatest challenge had been that of finding appropriate motive power. Without a satisfactory method of operating trains to and from New York, the extension could never have been built. The Pennsylvania would have had to continue suffering the humiliation of terminating on the west bank of the Hudson while the New York Central offered its riders the convenience of an intown station. Even if Lindenthal's bridge had been built in place of the tunnels, the problem of smoke and the subsequent legislative ban on steam locomotives would have forced the PRR to turn to electricity. In this respect, the New York extension of the Pennsylvania Railroad represented a new high in the advance of American technology. As

F. H. Shepard of Westinghouse commented, "It should be the cause of supreme satisfaction to all engineers to know that this gateway to New York City was predicated entirely upon the use of electricity."[51]

From a different perspective, the electrification of the New York extension symbolized not a technological breakthrough but a conservative embrace of proven technology. The extension became the last major rail electrification on the continent to use low-voltage direct current, for example. And the DD1, which perhaps did represent, as one Westinghouse engineer boasted, "survival of the fittest in railway evolution,"[52] nonetheless was in several ways a throwback to steam locomotive design. The perceptive *Scientific American* observed, "It is a curious instance of what might be called the vagaries of mechanical evolution that the latest and most powerful of electric locomotives . . . should be furnished with those side and connecting rods, the abolition of which the electric locomotive was considered to be one of the principal points of improvement."[53]

Whatever their mechanical heritage, the Pennsylvania was extremely pleased with the performance of the DD1's. These high-adhesion machines showed not the slightest strain when called upon to start 850-ton trains on the steep 1.93% grade westbound in the Hudson tunnels—a task which would have set the drivers of a steam locomotive spinning helplessly. The DD1's were equally at home rocketing across the Meadows towing 14-car limiteds at speeds of more than 80 miles per hour. So confident was the PRR in the ability of its new box cabs that it dared to equip its signal system in the tunnels to accommodate 26 movements per hour, an arrangement that allowed just over two minutes headway between trains.[54]

Now only 13 minutes were required for an 8.8-mile jaunt between Harrison Transfer (which the PRR soon renamed Manhattan Transfer) and Penn Station. This time compared very favorably to the 15 minutes needed for the ferry trip alone between Exchange Place and lower Manhattan. The railroad still encouraged commuters to disembark at Manhattan Transfer and use either the ferries or the H&M's trains to Manhattan. To add force to this wish, it slapped a ten-cent surcharge on all tickets routed to Penn Station and offered a discount ticket to patrons using Exchange Place. With only two tunnels under the Hudson, the Pennsylvania had enough long-haul trains to worry about without being bothered by a crush of marginally profitable local trains. The number of long-distance runs increased to the point where the railroad had to purchase nine more DD1's in 1911 to cope with the growing traffic.[55]

Beginning in 1917, passengers arriving at Pennsylvania Station were offered through service to and from New England. That year, the jointly owned (by PRR and New Haven) New York Connecting Railroad was opened for service between Port Morris and Sunnyside

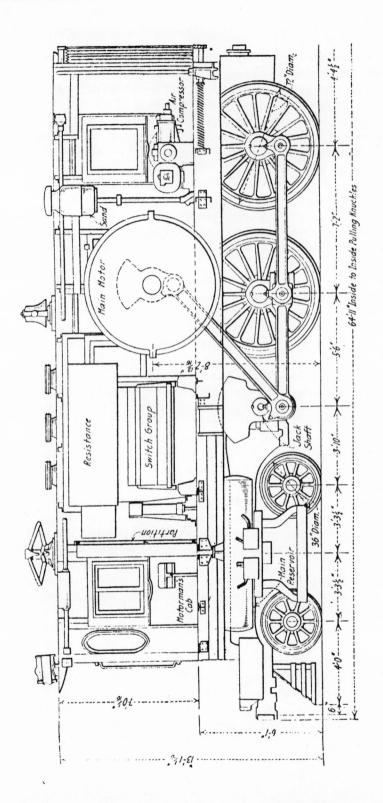

Cutaway view of a single DD1 unit. (*Street Railway Journal*)

via Lindenthal's colossal Hell Gate bridge. The New Haven erected a.c. catenary as far as Sunnyside and then ran over the Pennsylvania's third rail into Manhattan. The two railroads offered not only through car service to and from the west but also between Boston and Washington. The PRR and the New Haven likewise initiated through freight service to and from New England, using the Connecting and the Long Island railroads and the PRR's car-float operation across New York Bay.[56]

The Pennsylvania's electrification attracted much attention from railway men. It became a source of great pride to the railroad, especially as the DD1's tallied mile upon mile with a minimum of downtime. C. B. Keiser, master mechanic of the newly created Manhattan division (which encompassed the New York extension) informed the American Railway Master Mechanics Association that the 33 DD1's had operated a total of 994,592 miles in 1912. Detention time to passenger trains owing to locomotive malfunctions was a mere 66 minutes for that year—and only 38 of these minutes could be attributable to failures of an electrical nature. Repairs to the DD1's, Keiser stated, amounted to "practically nothing." The electric locomotive committee had made every effort when designing these machines to ensure accessibility of components. When a DD1 did suffer a breakdown, the enginemen themselves could often make the necessary repairs out on the road. For heavy repairs, the railroad sent the locomotives to its Meadows shop. There an entire cab could be lifted off in one piece to allow mechanics convenient access to its internal parts. The mechanical and electrical designs of the locomotives were so elementary that shopmen who never before attended to electric engines required only a few months' experience before they became experts at the entire cycle of repair and maintenance.[57]

Steam advocates at first attributed the excellent performance of the DD1's to the fact that they had been in service just a brief period of time. New equipment almost always gave a good account of itself in the early stages. As years passed, however, the sparkling record of the PRR electrics failed to dim. By November 1914, the DD1's had accumulated a total of 3,974,746 miles and had experienced only 45 failures. Five years later, locomotive mileage had climbed to 9,508,765, with the number of engine failures rising to 121, or 1 failure for every 78,600 miles. Even by 1921, maintenance costs for the electric fleet averaged about one-sixth of that for PRR steam locomotives.[58]

Westinghouse was no less ecstatic than the Pennsylvania about the DD1's accomplishments. In 1914, Westinghouse received permission from the railroad to display one of the locomotives at the Panama-Pacific Exposition to be held in San Francisco the following year. The exposition was a mammoth world's fair staged to commemorate the opening of the Panama Canal. At San Francisco, Westinghouse placed the DD1 in the Pennsylvania State Building, one of eight structures comprising the Palace of Transportation. The locomotive sat beneath the dome of the building on a revolving turntable. The public was invited to pass through the locomotive cab and inspect the internal machinery. Nearby, Westinghouse had created a 26-foot square diorama of the New York metropolitan area, which included scale models of skyscrapers, bridges, docks, and, of course, the Pennsylvania Tunnel and Terminal Railroad Company.[59]

A Westinghouse public relations man called the electric locomotive "the dominating feature of the entire Palace." The exposition's Jury of Awards concurred. It bestowed on the PRR machine the Grand Prize in Transportation in recognition of the design, construction, and near-perfect operating record of the DD1 fleet. The panel had no easy task in making its selection. It had to choose from a wide variety of vehicles, ranging from the latest oil-burning Mallet steam locomotive to a shiny new 12-cylinder automobile.[60]

At San Francisco, the DD1 overcame all hazards to triumph in the end, just as it had done five years before, several thousand miles to the east. It prevailed in spite of the fact that it did not represent much of a technological innovation. The locomotive not only embodied conservative mechanical design; the whole low-voltage d.c. system was soon rendered obsolete by advances in a.c. and high-

DD1 No. 14 (roster nos. 3960-61) at the PRR's Wilmington shops, 1939. The locomotive's appearance had changed little since its delivery to the railroad in 1910. (William E. Grant)

voltage d.c. systems. Nevertheless, the DD1 did provide the low-cost, dependable rail transportation that the Pennsylvania had been seeking for its New York terminal operation since the day Alexander Cassatt had decided to put his faith in electric traction.

3

....

Alternating Current
at Philadelphia

The tributes heaped upon the Pennsylvania Railroad in response to its New York improvements had hardly subsided when the railroad disclosed that it had another electrification project under study. President McCrea told the PRR's shareholders at their annual meeting in the spring of 1912 that the railroad was considering converting to electric operation several of its suburban lines radiating from Philadelphia's Broad Street Station. Passenger traffic to and from the station—the PRR's main terminal in the city—was extremely heavy. Electrifying the short-haul traffic on key routes might be a more economical way of relieving this congestion than increasing the size of the station, said McCrea.[1] In the final analysis, electrification did represent the best method of alleviating the overcrowded conditions. In its subsequent implementation of electric traction at Philadelphia, the Pennsylvania displayed the same dogged determination to develop equipment of the utmost reliability as it had at New York. On the other hand, at Philadelphia the railroad evidenced considerably more foresight in planning its system and a greater willingness to assume the role of a technological innovator.

All trains destined for Broad Street Station left the main line in West Philadelphia, crossed the Schuylkill River, and ran eastward for about a mile atop a series of trestles and stone arches popularly known as the "Chinese Wall." A short distance beyond the eastern end of this wall, directly across from City Hall, sat the station, built in 1881 and enlarged in 1893. The terminal yard was of the stub-end type, meaning that it lacked through tracks. Outbound trains had to exit the same way they had entered, picking their way through a maze of switches and battling hordes of inbound trains until they had safely recrossed the Schuylkill and rejoined the main line. In an effort to reduce the density of traffic, the Pennsylvania in 1903 constructed a new passenger station in West Philadelphia to serve most north-south through trains, thus eliminating the need for

Broad Street Station, a few years before its demolition. (Conrail)

them to cross the river and enter Broad Street Station. A year later, the railroad erected a station along its main line in North Philadelphia to replace Broad Street as the principal stop for east-west through trains.[2]

While these improvements did relieve overcrowding somewhat, Broad Street Station and its approaches remained clogged with traffic. All long-distance trains which originated or terminated in Philadelphia used the downtown station, as did semilocal runs to such cities as Harrisburg and Pottsville in Pennsylvania and Atlantic City in New Jersey. The prime cause of congestion, however, was the marked growth of traffic to and from the Philadelphia suburbs. In 1911, nearly 600 trains (about half of them suburban locals) used station facilities originally designed to accommodate only 160. One knowledgeable observer avowed that the terminal was the site of "what is undoubtedly the most severe congested traffic problem in modern railroading."[3]

To enlarge the station a second time would be difficult and costly.

The structure was located in the heart of downtown Philadelphia. The PRR would pay a high price for the additional real estate it needed, assuming, of course, that it could find property that was for sale in this heavily developed district. Even then, increasing the number of tracks in the terminal area would solve only part of the problem. The railroad would have to add more tracks to its main line through West Philadelphia and beyond to the suburbs. In the post-Civil War decades, the Pennsylvania played the part of land speculator and had vigorously promoted the growth of towns which lay near its tracks just west of the city line. Expansion had been gradual until near the close of the century, at which time thousands of Philadelphians began flocking to the suburbs to escape the increasingly unattractive conditions of life in the city. So many affluent families took up residence in such communities as Paoli, Bryn Mawr, Radnor, and Overbrook that the term "Philadelphia main line" came into use with reference to the social geography of the city's aristocracy.[4] The Pennsylvania now had to find a way to provide more efficient transport for the increasing number of people who worked downtown but resided elsewhere.

The only alternative to physically expanding Broad Street Station and adding more main-line tracks was converting at least some trains to electric propulsion. Electrically powered multiple-unit cars were able to accelerate much faster than those drawn by steam locomotives. Speedier getaways from the stations translated into reduced running times overall. If the movement of trains over a given segment of trackage could be quickened, more trains could be run. In other words, electrification allowed a railroad to augment the train-carrying capacity of a line without resorting to the necessity of laying additional tracks.

As early as 1894, the PRR had made some preliminary surveys for electrifying a portion of its local passenger service in and around Camden, New Jersey, just across the Delaware River from Philadelphia. In fact, the railroad went so far as to convert its 7-mile Burlington and Mount Holly (New Jersey) Branch to electric traction. Three wooden d.c.-powered combines, each pulling a nonmotorized coach, handled local passenger chores on this branch beginning in 1895 in what was one of the very first steam railroad electrifications in the United States.[5]

Each of the combines measured 35 feet in length (slightly shorter than the average PRR coach of that era) and featured an operator's compartment and a small baggage space at one end. The remainder of the car was devoted to passenger seating. Jackson and Sharp of Wilmington, Delaware, manufactured the cars, while Westinghouse supplied the traction motors, all of which were of the axle-hung variety. The railroad built its own power plant at Mount Holly. There a Westinghouse steam-driven turbogenerator furnished direct current at 500 volts. Trolley poles mounted on top of the cars collected the current from a contact wire which hung suspended from wooden poles placed at 110-foot intervals along the right-of-way.[6] From the standpoint of equipment and operation, the Burlington and Mount Holly Traction Railroad Company could hardly be distinguished from an interurban line. Had not the powerful Pennsylvania Railroad been its parent, the trade press most likely would have made little note of its conversion to electricity.

The branch was ideally suited for an experiment with electric traction. Local passenger trains constituted the bulk of its traffic. Topographic and operational conditions closely resembled those of the Camden area. More importantly, the Burlington and Mount Holly had been a steady money loser for years. If electrification could attract enough new riders to reverse its financial fortunes, the PRR was prepared to electrify other unprofitable branch lines, first at Camden and eventually at Philadelphia. The significance which the railroad attached to this project was revealed on June 3, 1895, a few weeks before the PRR offered electric service to the general public. On that day, President Roberts and several other members of the corporate hierarchy spent a couple of hours riding to and fro over the line in one of the new Jackson and Sharp cars.[7] The Pennsylvania did not lavish such executive attention on obscure spur lines like the Burlington and Mount Holly unless it had great expectations.

In the 1890s, the PRR viewed electrification primarily as a cost-cutting tool. Suburban and intracity business was not very profitable for steam railroads. Hauls were not long enough to reimburse the railroads for their sizable investments in rolling stock and physical plant. In addition, new competitors, the electric street railways, threatened to draw riders away and compound the steam roads' losses. *Railway Age* estimated in 1895 that on some routes in the Philadelphia area the PRR and the Reading (the city's other major carrier of local passengers) had lost more than 50% of their business during the past five years to street railways. As patronage fell, operating costs per passenger-mile rose. The Reading cut the frequency of its trains in an attempt to avoid further losses.[8] The Pennsylvania sought a different solution by trying to reduce the costs of operation, hence the Burlington and Mount Holly experiment.

Electrification brought to this branch line a definite improvement over steam service. A one-way trip between Burlington and Mount Holly now required just 20 minutes, bettering the former steam schedule by 5 minutes. The railroad reduced the one-way fare by a nickel (to ten cents) and dispatched ten electrically propelled locals in each direction six days a week, as opposed to the six trips pre-

Burlington and Mt. Holly No. 3, an electrically powered combine. (Altoona Public Library)

viously provided by steam trains.[9] The handful of through passenger trains and freight trains which used the double-track line continued to be hauled by steam. Late in 1901, this state of affairs changed radically when a fire gutted the Mount Holly powerhouse. A few days after the fire, the Pennsylvania announced that it would not rebuild the plant. The line would henceforth revert to complete steam operation. For the record, railroad spokesmen declined to give any reason for this course of action. They admitted privately, however, that this first experiment with electric traction had not fulfilled the company's hopes. The electrical equipment had functioned almost flawlessly. The PRR had no complaints on that score. Passenger revenues had risen, also. Still, the railroad did not deem that increase sufficient to warrant spending the relatively large sum of money which rebuilding the powerhouse would entail.[10] Nor had the merits of electrification in general impressed the Pennsylvania to the point where it considered further experiments desirable. Down came the curtain on the railroad's first engagement with electric traction.

Now, in 1911, the PRR was once again pondering the use of electric motive power. But times had changed since the 1890s, and so had the reasons for the railroad's interest in converting to electricity. Suburban passenger business, just beginning to swell in 1895, had reached flood tide 15 years later. Reducing operating expenses or attracting even more riders did not matter nearly so much anymore. The PRR concerned itself now primarily with boosting the ability of its stations and tracks to accommodate the throngs of people travelling to and from the suburbs. It had to act quickly, before the local traffic grew to such proportions that it interfered with the more profitable long-haul passenger and freight business.

To assist the board of directors in making a decision, President McCrea called in George Gibbs, fresh from the New York tunnel project, to study the feasibility of electrifying some of the suburban routes. Gibbs had recently joined with Earnest Rowland Hill, his assistant during much of the extension work, to form the consulting firm of Gibbs and Hill. As specialists in electric railway engineering, the two engineers and their company were to exercise a profound and long-lasting influence on the course of electric traction, not only on the Pennsylvania, but on several other large eastern lines as well.[11]

Late in 1912, Gibbs reported to the directors that they had little choice but to approve a conversion of some routes to electric traction. Enlarging Broad Street Station (assuming the land could be acquired) involved costs producing no direct return on the investment. Electrification, by contrast, would reduce operating expenses substantially, although probably not enough over the short term to cover the initial installation costs. Gibbs further pointed out that the PRR was powerless to alter the stub-end nature of the station yard, even if it could add several more tracks. Therefore, the railroad could not eliminate most of the switching done at the terminal—turning the locomotive and transferring it to the other end of the train, for example. If, however, the PRR introduced electric multiple-unit car service, there would be no reason to turn the locomotive and run it to the other end, since there would be no locomotive in the first place. The railroad could also do away with the expensive, time-consuming procedures of taking on water and coal.[12]

The Pennsylvania did not have to be sold on the value of m.u. cars. The successes achieved by these cars in heavy-duty service on the New York Central, the New Haven, and the PRR's own Long Island did not go unnoticed in the Philadelphia general offices. Thus when the Pennsylvania's engineers in 1910 designed a new, all-steel coach to be used system wide in steam-powered suburban service, they wisely made certain provisions which would permit the railroad to convert the cars to electric propulsion easily and economically. The new coach, Class P54, was 64 feet long and had a reinforced underframe which could support the heavy electrical gear with no further strengthening. All that had to be done to modify the P54's for m.u. duties was to install a motorman's compartment at either end and substitute a motorized truck for a nonmotorized one.[13]

Gibbs and Hill's assessment elicited a favorable response from the PRR. Nevertheless, before the railroad gave electrification its official blessing, it had to confront the perennial a.c. versus d.c. issue.

Neither the Pennsylvania nor the consultants were inclined to accept direct current solely on the basis of its prior use in the New York area. George Gibbs himself had said in 1910 that the reasons the PRR utilized direct current in New York "apply only to this special installation. No broad generalization is intended for other traction projects, where local conditions must govern."[14]

The railroad had been considering for several years the possibility of a long-distance electrification project. All four primary operating divisions between New York and Pittsburgh had come under scrutiny as candidates for conversion to electric traction. If the Pennsylvania eventually chose to electrify one or more of these divisions, probably it would use alternating current. The New Haven had already proven the practicability of the a.c. system. Indeed, the road was so pleased with alternating current that it was extending catenary another 60 miles eastward to New Haven, Connecticut. Logic dictated that the Philadelphia suburban electrification be compatible with the long-distance electrification, which at some future date would probably encompass the Philadelphia installation. With a view toward a major extension of electric operation at some not too distant time, then, the Pennsylvania's directors on March 12, 1913, authorized the implementation of an alternating current system from Broad Street Station to Paoli, 20 miles distant.[15]

A few preliminary hurdles had to be cleared before construction could begin. First, the PRR had no experience with alternating current except as employed in testing locomotive No. 10,003. In order to increase their familiarity with the mechanical attributes of a.c. distribution equipment, the railroad's engineers erected a mile of unenergized catenary spanning the four-track main line between Radnor and St. Davids. Steam locomotives puffed back and forth beneath the wires pulling work cars on which several different kinds of pantographs had been mounted. This allowed PRR and Gibbs and Hill engineers to achieve the best possible collection and catenary designs.[16]

The second major problem that the railroad faced concerned the selection of a source of electric power. In both the Burlington and Mount Holly and the New York electrifications, the Pennsylvania generated its own electricity. In 1911, all steam railroad electrifications save one depended upon power generated exclusively in railroad-owned plants.[17] Steam roads had not yet seriously explored the notion of purchasing commercially generated current. Rail operations tended to be cyclical. A great many trains might run for a short period of time, say, during the morning rush hour; but traffic would be relatively light for the rest of the day. Then came the evening rush hour and traffic would crest again for a brief time. Cycles varied according to railroad and location, as well as time of day. On roads dispatching many passenger trains, traffic levels could be depended upon to rise and fall with clocklike regularity. Freight traffic was more unpredictable. The number of freight trains dispatched depended less on printed schedules than on such highly variable factors as availability of motive power, crews, and cars, as well as on the state of the local and national economies. Electric railroads therefore demanded tremendous amounts of power at certain times of the day or week or even year and drew fairly small quantities during other periods. Given the limited amounts of excess power that most commercial utilities could supply in this early stage of their development, railroads converting to electric traction were understandably hesitant to rely on them. The railroads preferred to build their own generating plants. In this way, they could be assured of having a sufficient reserve of current whenever they needed it.[18]

Utilities at this time also limited themselves to serving local markets. No interconnecting grid system existed to allow companies to buy and sell electricity to one another. Practically every community had its own power company. Conversely, railroads spanned great distances and transcended local markets. A railroad preparing for anything more than an electrification of a few miles had to contend with the unattractive task of negotiating separate contracts with a multitude of utilities.[19]

For their part, the commercial power producers were equally cool toward the prospect of signing railroads as customers. Power companies feared the unbalanced load of the electric railway. The peak demands of a railroad could rob other customers of an adequate supply of current. If utility personnel did not act promptly in such a situation, the resultant power shortage could damage customer-owned appliances and even harm equipment at the central generating station. Utilities could furnish enough electricity for both rail and nonrail consumers only by vastly augmenting their generation and distribution capabilities. In these pre-World War I years, unfortunately, most commercial power companies experienced a shortage of investment capital. What money they could attract in most cases just barely enabled them to meet the needs of the booming residential demand for electricity.[20]

Technical problems intensified the mutual antipathy between railroads and electric companies. The power industry had not yet settled on common standards regarding the characteristics of the current that they produced. Within any given region of the country, companies could be found that generated direct current at any one of a half dozen different voltage levels. Utilities supplying alternating current generated it at single-, double-, or triple-phase, at cycles ranging from a low of 15 to a high of 150.[21] How could a railroad that traversed dozens or even hundreds of commercial power markets ever hope to purchase an identical kind of current from one end of the line to the other?

Utilities in the Philadelphia area reflected the chaos prevailing in the industry during the 1890s. Twenty-six companies generated either direct current (at 110, 220, or 600 volts) or single- or polyphase alternating current at 60 or 133 cycles.[22] Yet the changes which occurred in the commercial power business in Philadelphia during the first decade of the twentieth century typified the changes which gradually transformed the industry as a whole. They ultimately led to the Pennsylvania Railroad becoming the first large American steam railroad to sign a purchase of service agreement with a commercial utility for motive power purposes.

Within a few years of its incorporation in 1899, the Philadelphia Electric Company had absorbed the myriad of small firms in the area. Centralized ownership and management brought about the badly needed standardization. Philadelphia Electric continued to supply old customers with direct current, for example, but it provided all new users with alternating current. The company decided to generate nothing but three-phase, 60-cycle power for a.c. customers, since this type of current could be utilized by industrial and residential customers alike and was thus fast becoming the national norm. Philadelphia Electric made an exception to this standard in 1911, however, when it contracted to furnish the Philadelphia Rapid Transit Company with 25-cycle alternating current. The transit company had just reorganized in an attempt to shed its traditional image as a political football and model of inefficiency. Its new management began to shut down the company's own generating stations on the grounds that current could be purchased at a much cheaper rate.[23]

The Pennsylvania Railroad carefully noted the terms of the transit company's agreement with Philadelphia Electric. After comparing the cost of commercial power to the costs of self-generated power, as estimated by Gibbs and Hill, the railroad discovered that the advantage clearly lay with commercial power. Electric company and PRR representatives opened formal talks in June 1913 aimed at formulating a mutually satisfactory power contract for the Philadelphia-Paoli electrification. The utility was already planning a substantial enlargement of its Christian Street generating station, which served downtown Philadelphia and which would supply power to the railroad if a contract could be arranged. A slight further expansion of the station to ensure sufficient current for the PRR would not be unduly expensive. By 1914, Philadelphia Electric had acquired many large industrial customers and had to increase its generating capacity accordingly. The railroad demand, therefore, did not represent the same troublesome peak that it had just a few years earlier. The utility had added to its generating facilities in order to serve many heavy users, not just one. It could thus spread the cost of installing additional equipment among all these industrial consumers. No one customer would have to bear the burden of paying for plant expansion.[24]

In 1914, the railroad and the electric company signed a five-year contract in which the utility pledged to supply power not only for the Paoli electrification but also for an extension to Chestnut Hill. The PRR's directors could not have anticipated much difficulty in coming to terms with Philadelphia Electric, for they voted to electrify the Chestnut Hill Branch in September 1913, several months before the power contract was agreed upon. Trains to and from Chestnut Hill shared the main line with Paoli-bound traffic as far as Mantua Junction, about a mile from Broad Street Station. At this junction, Chestnut Hill trains turned northeast and followed the New York division main line for another mile to the North Philadelphia Station. There the Chestnut Hill line veered northward and eventually crossed the city line. A total of 11 route-miles were to be electrified, but work would not begin until the Paoli line had been completed.[25]

The power contract represented a milestone in the growth of both signatories. Never again would the Pennsylvania resort to generating its own electricity. In fact, a few years later the railroad reached an agreement with Philadelphia Electric whereby the PRR-owned West Jersey and Seashore Railroad bought power from the utility in place of current formerly generated by the Seashore's Westville, New Jersey, plant. And thanks to what Philadelphia Electric Vice-President W. C. L. Eglin described as the PRR's "safe, sane, and progressive policy," his company would assume in time a larger railroad electrification load than any other utility in the world.[26]

Erection of catenary commenced in the spring of 1914. Railroad employees did most of the work, under the supervision of railroad engineers and those of Gibbs and Hill. Based on the results of the experimental Radnor-St. Davids line, the PRR selected a tubular cross-catenary distribution system. Tubular steel poles set in concrete foundations were positioned at 350-foot intervals on either side of the four-track main line. Two heavy-gauge wires running at right angles to the tracks connected each pair of poles. A messenger wire running above each of the tracks was attached to these cross wires. Adopting a technique pioneered by the New Haven, The Pennsylvania's engineers suspended from each messenger a lighter wire, or auxiliary messenger, whose purpose was to lend a consistent stiffness to the longitudinal wiring. It also made sure that the contact or trolley wire, which was held by a series of clips several inches below the auxiliary messenger, hung at a uniform level above the rails. Pantographs, unlike the old trolley poles used on street railway cars, could not readily adjust themselves to variations in height of the contact wire; thus a precise distance had to be maintained at all times between the top of the rail and the contact wire in order to ensure a continuous electrical circuit.[27]

At the peak of construction, the railroad had over 1,000 men on the job. Only about half were directly engaged in electrification work, however. The PRR had decided to undertake other major improvements to the Paoli line in order to take maximum advantage of electrification's potential. The railroad did away with all grade crossings between Broad Street and Paoli, thereby removing the danger of a crossing accident involving a speeding m.u. train. Accelerated schedules were now possible also, since there was no longer the need to reduce speed at certain highway crossings. The railroad's implementation of a new signal system figured in the quest for more speed, too. Formerly, the Pennsylvania had used the same combination of semaphores and colored lights here as it had on most of its other main routes. The new system used position lights for both day and night. The position of a row of five white lights set against a black background now governed train movements. A vertical position indicated proceed, for example, and horizontal, stop. The new signals, a novelty on the American railroad scene, could be spotted much more easily than semaphores amid the jumble of overhead wires. The intensity of the lights was such that train crews could read the signals in broad daylight at distances of nearly a mile.[28] Position lights proved so satisfactory that within a few years the Pennsylvania adopted them for system-wide use.

Following the precedent that it had established at New York, the PRR insisted that every employee whose job in any way related to electrification be thoroughly trained in his duties. The railroad procured two P54 coaches, equipped them with electrical gear similar to that which it was placing in the m.u. cars, and converted the interiors to classrooms. One car was sent to Broad Street Station, the other to the West Philadelphia shops. All employees who aspired to enter the new electrified service had to familiarize themselves with the operation of these test cars. After passing written and oral tests on the stationary cars, would-be trainmen gained more experience and had to pass yet another series of examinations on a part of the route that had already been electrified (but not opened for service). To aid its employees in meeting these stringent requirements, the Pennsylvania Railroad branch of the Young Men's Christian Association, in association with the extension division of the Pennsylvania State College's School of Engineering, offered a 20-lesson course in electric railway operation to all interested personnel, no matter what their present position with the railroad. About 300 men signed up for the classes. Approximately 100 of these employees currently worked as enginemen or firemen in suburban steam service.[29] The impending electrification threatened to eliminate few, if any, jobs, however. The engineman and fireman would be replaced by a single motorman, of course, but otherwise the crew size of most suburban trains would not be affected. In fact, the increased frequency of trains permitted by multiple-unit operation might increase the need for personnel. And the handful of employees who might be furloughed could easily be shifted elsewhere, for the scope of steam operations in and around Philadelphia was expanding (owing to increased long-haul passenger and freight business) rather than shrinking.

The only problem of any consequence which the PRR experienced pertained to its need for 25-cycle, single-phase current. Philadelphia Electric generated most of its alternating current at 60 cycles, three-phase. The 25-cycle, single-phase electricity which it produced for the transit company was not sufficient to be shared with the railroad. For technical reasons, American railway electrifications using alternating current had by now standardized on 25 cycles. The lower the frequency, the smaller the series-commutator motor could be for a given horsepower output; that is, a motor of a particular size and weight operating at 25 cycles had a higher horsepower rating than an identical one operating at 60 cycles. As the frequency increased, so did the dimensions of the motor, in order to maintain a specified horsepower. This relationship was of little importance for home or factory use. For railroads, it was vital. An electric locomotive had to pack as powerful a motor as possible into a very restricted space. In addition, railways using low-voltage d.c. systems but relying on an a.c. generating and distribution system found that the lower the frequency of the alternating current, the more efficiently it could be converted to direct current.[30] Consequently, even such d.c.-equipped lines as the New York Central and the Long Island generated alternating current at 25 cycles.

Railroads differed from most other industrial consumers of electricity in still another respect. With one exception, American a.c.-powered railways operated with single-phase current, rather than the three-phase current which most utilities produced. Three-phase power, although popular among European railroads, required an extremely complicated distribution system. Usually the running rails themselves served as the conductor for one phase, while two overhead wires carried the other two phases. These wires naturally had to be insulated from one another. At turnouts and crossings, this became a real problem. Difficulty in obtaining the proper degree of insulation of the two overhead wires limited the voltage level at which the system operated. The Great Northern Railway energized its new 4-mile Cascade Tunnel electrification in 1909 at 6,600 volts. It was destined to be North America's only three-phase steam railroad electrification.[31] On the other hand, single-phase installations needed only one overhead wire, which posed no insulation troubles and hence could withstand a much higher voltage. Most single-phase electrifications operated at 11,000 volts.

The Philadelphia Electric Company already produced 25-cycle current for the transit line and a few industrial customers. It could — and did — buy additional 25-cycle generators to meet the PRR's demands. Practically all of the a.c. power generated by Philadelphia Electric, however, was of the three-phase type. Supplying a large single-phase load from three-phase equipment would throw the entire distribution system of the utility off balance and cause serious damage. A way had to be found to spread the single-phase load of the railroad equally and automatically across all three phases of the utility-generated current.[32] Electric company and PRR engineers had been working to resolve this difficulty since early in 1914. As the summer of 1915 approached, they had still not developed a practical means of balancing the load. Meanwhile, work continued on the catenary system and other improvements to the Paoli line. The Pennsylvania had publicly promised to begin electric service sometime in the fall. Officials in the railroad's general offices on the upper floors of Broad Street Station must have watched their calendars with some discomfort as the days and weeks passed with no apparent progress being made on the phase-balancing problem.

The management at Philadelphia Electric was no less uneasy. In desperation, President Joseph B. McCall dispatched his two top engineering officers, W. C. L. Eglin and Horace P. Liverslidge, to the headquarters of the General Electric Company at Schenectady. General Electric had recently introduced a new type of phase balancer thought to be capable of handling the heavy loads contemplated by Philadelphia Electric. But because they had not yet had the opportunity to subject it to a thorough battery of tests, GE engineers refused to say whether the balancer would function properly for extended periods of time. The dilemma facing Eglin and Liverslidge can well be imagined. Should the electric company purchase and install the apparatus and run the risk of damaging its whole system? Or should it await the outcome of further trials with the machine before buying it, a course of action which guaranteed public embarrassment and might even bring on a lawsuit from the railroad? At length, GE called in its highest-ranking consulting engineer, Charles P. Steinmetz, one of the world's foremost authorities on the theory and design of electrical equipment. Steinmetz spent less than half an hour examining the balancer and making a few mathematical calculations, after which he affirmed laconically, "It will work." Primarily on the strength of these three words, Philadelphia Electric immediately bought two of the huge machines and installed them in its Christian Street Station. There Steinmetz's calculations were vindicated. The balancers performed faultlessly. The PRR could have all the single-phase current it wanted.[33]

The utility had just put into operation at the same facility the world's largest pair of steam turbogenerators. One unit, rated at 30,000 kilowatts, produced 25-cycle, three-phase current. The other carried a 35,000-kilowatt rating and churned out three-phase alternating current at 60 cycles. Seven older and smaller units gave Christian Street a total generating capacity of 86,000 kilowatts, making it the most powerful central station in the country. After having been converted to single-phase, current destined for the PRR was transmitted at 13,200 volts through submarine cables beneath the Schuylkill River to the railroad-built substation at Arsenal Bridge in West Philadelphia. There the level was stepped up to 44,000 volts to enhance the current's transmission efficiency. From Arsenal Bridge the electricity flowed to four more railroad-owned substations: West Philadelphia, Bryn Mawr, Paoli, and Chestnut Hill. Transformers at each of these locations stepped down the current to 11,000 volts, at which level it was then fed into the contact wires. The entire distribution system required the presence of just one person—a power director at the West Philadelphia substation. This individual in effect performed many of the same functions for the m.u. cars as firemen performed on steam locomotives. In emergency situations, the equipment in the other four substations could be remotely controlled by men in nearby interlocking towers.[34]

While wire crews were completing the last of the electrical work, the railroad began gathering its m.u. fleet at Paoli, where it had built a new yard to store and maintain the cars. The PRR had gradually removed 93 P54 cars from steam service and sent them to the Altoona shops. There workmen removed one of the trucks and in its place installed a powered truck containing two 255-horsepower Westinghouse traction motors, which were the largest and most powerful that the railroad could fit between the wheels of these cars. After the addition of a motorman's compartment at each end and a transformer to step down the line voltage, the cars were given a new coat of the standard tuscan red paint (with gold lettering) and a new class (MP54) and were finally shipped east again to undergo inspection and testing.[35]

The Pennsylvania energized all 20 route-miles between Broad Street and Paoli for the first time on September 4, 1915. During the following week, all the MP54's ran empty over the line so that the railroad inspectors could detect any flaws which heretofore might have gone unnoticed. As in the case of the grand opening of Manhattan's Pennsylvania Station five years earlier, the railroad chose to begin electrified service on a weekend. In all likelihood, the decision resulted from the PRR's desire to correct any deficiencies that might

MP54 No. 626, freshly outshopped for the PRR by the Standard Steel Car Company in 1926. (Altoona Public Library)

suddenly appear in time for the MP54's to bear the brunt of the large crowd of weekday patrons. Thus on Saturday, September 11, regular revenue service began as a group of four of the new cars departed from Paoli at 5:55 A.M. for the day's first run to Broad Street Station. Despite the earliness of the hour and much to the railroad's surprise, thousands of spectators in anticipation of this momentous event lined the four-track right-of-way all the way to the terminal. As the m.u.'s whirred away from the station platform at Paoli, a hearty cheer burst from the crowd, a salute which kept pace with the train for the remainder of its journey. At first, the Pennsylvania carded only one electrically powered round trip daily on this route. By mid-October, however, electricity had banished steam from every one of the 68 locals scheduled on the Philadelphia-Paoli run.[36]

Initially, the railroad ran its m.u. trains to coincide with the former steam timetables. The MP54's had no trouble adhering to this schedule, owing to their rapid acceleration. At year's end, the

PRR reduced the eastbound running times between Paoli and Broad Street from 59 to 49 minutes, and still the electric cars had no problems keeping time. The Pennsylvania's foreman of electric car equipment, L. M. Willson, reported that after two years of service the MP54's were turning in an on-time performance better than 90% of the time, compared to an 80% on-time record established by steam-drawn locals. A joint Westinghouse-PRR survey taken in 1921 revealed that the m.u.'s were still running "on the advertised" 94% of the time. The same study showed that during an average month the MP54 fleet accumulated about 250,000 miles with but five detentions totaling just 14 minutes.[37]

The railroad had spent $4.2 million to electrify and otherwise improve its main line to Paoli. This same amount of money would have bought and built a single track right-of-way 400 miles long in most parts of the country. Yet the Pennsylvania judged that it had made a sound investment in upgrading this 20-route-mile, 93-track-mile suburban line. As soon as railroad officials recognized the degree to which electrification improved service, they turned their attention to the Chestnut Hill Branch. Work had hardly begun there when the United States became involved in World War I. War-related shortages of skilled laborers and materials so delayed construction that the m.u. fleet did not begin running to and from Chestnut Hill until early in April 1918. The PRR's schedule called for 21 electric trains per day in each direction on this mostly double-track route.[38]

A number of safety and operational problems which the railroad had not experienced with steam accompanied the switch to electric traction. The Pennsylvania had foreseen a few of these difficulties and had taken steps to prevent their occurrence. Since the MP54's needed only one operator, the railroad incorporated a "dead man's" device into each car. This mechanism caused the current to be interrupted and the car to come to a halt in the event the motorman became incapacitated at the controls. To reduce the chance of accidental electrocution at the contact wire, the railroad placed a safety bar at the top of the ladder leading to the roof of each car. A man could not gain access to the roof unless he turned the bar away from the ladder. In doing so, he simultaneously lowered the pantograph, breaking contact with the 11,000-volt overhead line.[39]

Some problems the PRR did not spot in advance. The catenary quickly became the object of attention from small boys who delighted in testing one another's skills by shinning up the steel poles. Children's kites had an annoying habit of wandering toward the high-voltage wires and risking a nasty short circuit for the railroad as well as serious injury and even death to the youngsters. This situation became so grave that PRR representatives began to make regular visits to schools in neighborhoods through which the electrified tracks passed, warning the children of the hazards of playing near railroad property.[40]

Other problems were of a more technical nature. The exhausts of steam locomotives caused much of the main messenger wiring to corrode badly after only a few years' exposure. Maintenance crews had to replace nearly all the 1/2-inch galvanized steel messengers with 5/8-inch bronze wire, which withstood the effects of smoke and gas much more satisfactorily. In a similar fashion, the contact shoes on the pantographs of the MP54's wore down the thin copper contact wire much too rapidly, forcing the PRR to restring all its lines with heavier gauge wire.[41] Neither wires nor MP54's held up well during the first few sleet and ice storms to hit the Philadelphia area. Ice formed on the pantographs of cars stored outside the Paoli carbarn overnight and weighed on them so heavily that they could not maintain the proper spring tension to stay in contact with the trolley wire overhead. The ice could be melted only by shooting steam at the pantographs from hoses connected to steam locomotives. Ice also accumulated on the contact wire itself, forcing the railroad to run a shuttle of m.u.'s continuously over the electrified lines whenever severe storms struck. The pantographs were not able to keep the messengers free from ice, of course, and occasionally these wires did come down.[42]

By far the most spectacular plague which electrification visited on the Pennsylvania occurred at its five substations. Each of these two-story, brick and concrete buildings contained circuit breakers designed to shut off the current flow in the event of a short circuit or other fault in the line, such as a lightning strike. The breakers required a half second to trip open, however, which frequently was time enough for the current to race through and instantly wreck the substation with a blinding blue flash. Until the advent of high-speed (1/25 second) circuit breakers in the 1920s, the railroad was helpless to do anything to correct this troublesome state of affairs. About all it could do was to place rope ladders on the upper floors of the substations. Any employees trapped there when a short circuit occurred could simply throw the ladders out the windows and descend to safety. The public also suffered from these inferior breakers. Every time one failed to open fast enough, the resulting surge of current produced enough interference on adjacent telephone lines to cause every telephone within a 2-mile radius to ring furiously. An awful screech greeted anyone unlucky enough to be on the receiver at the time.[43]

Compared to the successes of electrification, all of the previously mentioned problems were mere nuisances for the railroad. George Gibbs estimated that converting Paoli and Chestnut Hill local service to electric traction increased the capacity of these two routes by 8%

MP54 No. 599 at Philadelphia, 1959, in the twilight of its career. (Harold K. Vollrath Collection)

and had the same effect as adding two more tracks beneath the mammoth train shed of Broad Street Station. Patronage had increased markedly since electric trains began running. Civic leaders, local residents, and the traveling public praised the Pennsylvania for the cleanliness of its new service. The smoke problem had never haunted the PRR in Philadelphia as it had the Long Island and the New York Central, but the railroad was nonetheless very pleased that so many citizens considered electrification to have been in the community's best interests.[44]

George Gibbs, while pleased with the success of the Paoli electrification, warned that it solved only part of the Pennsylvania's suburban traffic problem. He predicted that increased business on non-electrified commuter arteries over the next seven or eight years would again cause intolerable congestion in and around the Broad Street terminal.[45] Fortunately, the Paoli electrification pointed to a solution. By converting other suburban routes to electric traction—a course of action now shown to be eminently feasible—the railroad could readily cope with the expected growth in ridership.

4

....

The Electrification
That Might Have Been

Since steam engines simply could not operate through long sub-aqueous tunnels, electric traction was the only suitable motive power then available for the Pennsylvania Railroad's New York extension. The railroad opted for electrification at Philadelphia because, for various reasons, it could not enlarge its terminal facilities there. In both instances, electricity helped the PRR solve unique operating problems. Electric locomotives could do things at New York and Philadelphia that steam locomotives could not do. Electricity had not, however, conquered steam purely on the basis of operational and economic superiority in a situation unaffected by special conditions. Thus steam remained in command of the railroad's long-distance runs.

Only one railroad—the New Haven—had yet dared to undertake a lengthy electrification; but even with the extension to New Haven, that railroad had only 73 route-miles under wire. Most of its freights still ran behind steam.[1] The Pennsylvania contemplated electrifying on a far broader scale: an energized segment of 200 miles or more involving through traffic of all types. The PRR's adoption of alternating current for its line to Paoli demonstrated its confidence in such an ambitious project. Nevertheless, the railroad did not include the Philadelphia area in that portion of the system most seriously under consideration for long-haul electrification. Instead the Broad Street strategists had targeted the two divisions between Harrisburg and Pittsburgh as the most eligible candidates for conversion to electric traction. Traffic was extremely heavy on this section of the main line, and topographical conditions were as formidable as any encountered by a railroad east of the Mississippi—the route traversed the heart of the rugged Allegheny Mountains. Even prior to the Paoli electrification, the Pennsylvania's engineers had been studying the possibility of piercing this troublesome barrier with either third rail or catenary.

The Pennsylvania did not wholly lack experience with a long stretch of electrified line at this time. Electric trains had been running between Camden and Atlantic City on the West Jersey and Seashore subsidiary since 1906. Insofar as the Seashore's only electric rolling stock consisted of m.u. cars powered by low-voltage direct current, its impact on the PRR's plans for future electrification was minimal. On the other hand, the successful use of electric traction on the WJ&S did forewarn proponents of the steam locomotive that their machine might not be as invincible as they liked to believe.

The West Jersey and Seashore Railroad was the offspring of the 1895 union of the West Jersey and the Camden and Atlantic railroads, both PRR-controlled properties linking Camden and Atlantic City. The West Jersey had also run to Cape May. At Newfield, 30 miles beyond Camden, its main stem forked, one track continuing southeastward to Cape May, the other turning due east and heading to Atlantic City.[2] It was the line to Atlantic City that the PRR decided to electrify.

The railroad faced stiff competition from the Reading-owned Philadelphia and Atlantic City Railroad for the lucrative seasonal passenger business to Atlantic City and other nearby shore resorts. In recent years, the Reading had much the better of the rivalry. Its trains, drawn by new, big-boilered 4–4–2's, easily outpaced the PRR's trains, most of which were pulled by the smaller 4–4–0's. The Pennsylvania's desire to best the Reading, however, did not play much of a role in its determination to electrify. In this instance, the PRR much preferred to meet the challenge with an improved steam engine. Consequently, the Pennsylvania's chief mechanical engineer, Axle S. Vogt, designed his road's very first 4–4–2's with the Camden-Atlantic City service in mind. In 1899, Altoona outshopped a trio of these high-drivered machines, Class E1. They performed most commendably and could fly across the tabletop Jersey landscape with 300-ton trains at 75 miles per hour. These and subsequent subclasses E1a, E2, and E3 did everything their owners asked of them in competing with the Reading.[3]

The E's were the mainstays of the seashore run via Winslow Junction. This, the former Camden and Atlantic main line, was a good 15 miles shorter than the old West Jersey route through Newfield. The fact that the Pennsylvania elected to convert the longer of the two lines to electric traction and retain steam on the other is further evidence that the railroad did not have much interest in using electrification as a tool to beat the Reading. The PRR's electric trains running between Camden and Atlantic City in effect competed with the company's own steam trains. This intramural contest was precisely the kind the Pennsylvania wanted.

In the fall of 1905, A. J. Cassatt appointed a four-man committee to investigate the advisability of electrifying one of the routes to Atlantic City. The group consisted of a chairman, Fourth Vice-President Samuel Rea, and three other members: Lines East General Manager W. W. Atterbury, Chief of Motive Power Theodore N. Ely, and Alfred Gibbs. The Pennsylvania's management desired to compare the economies of steam and electric traction under nearly identical operating circumstances. What better arena existed in which to stage a confrontation between the two forms of motive power than the dual lines to the New Jersey shore? Atlantic City and other shore points were reaching their zeniths as America's favorite playgrounds, and traffic was heavy on both routes from May to September. The railroad wished to see if electric propulsion could cope with such congestion more efficiently than steam. In addition, the railroad planned to gauge the traffic-building potential of electrification. Most of the trains which ran via Winslow Junction were expresses, often carrying cars originating at distant points on the system. The Pennsylvania saw little sense in electrifying a line that generated scant local business. The longer route via Newfield carried considerably more local traffic, and so was the logical choice for initial electrification.[4]

Since the PRR had already chosen to electrify the Newfield line before the Rea committee was established, the primary task of Rea and his associates was to settle upon the technological aspects of the forthcoming transition from steam to electricity. They could recommend a single-phase alternating current installation, which many observers regarded as the system of the future, or they could suggest a less efficient but more reliable low-voltage d.c. system. After carefully weighing their options, the experts urged the adoption of direct current at 650 volts. Cassatt and the directors wanted to see some tangible results from electrification as soon as possible. Direct current equipment could be installed more quickly and cheaply than a.c. equipment and required no lengthy "break-in" period to eliminate hidden flaws.[5] The committee submitted its report in November 1905. A few weeks later, Pennsylvania officials announced the awarding of a contract for all electrical gear needed for the venture to General Electric.

In acknowledging the contract, GE boasted that the WJ&S electrification "marks another milestone passed in the march of progress of electric traction."[6] The statement was true enough in that the project would significantly increase the total electrified track-mileage of American railroads. General Electric treaded on thin ice, however, in terming this adoption of low-voltage direct current "progressive." It no more represented progress than did a similar decision three years later to embrace a 650-volt system for the New York tunnel extension. As it was to do on several important occasions in the future, the

Pennsylvania passed over the opportunity to become an innovator in the electric railway field in favor of remaining true to proven technologies.

Construction began early in 1906 under the supervision of the PRR's traction chief, George Gibbs. The railroad double-tracked the whole line with the exception of the heavily traveled 8 miles between Camden and Woodbury, which got three tracks. An overrunning third rail, identical to the 100-pound running rails, was used for most of the route. Because numerous highways intersected the right-of-way in the Camden area, the railroad installed trolley wire for the 4.4 miles between Haddon Avenue in Camden and the village of South Gloucester. In order to attract more local riders, the PRR also electrified the Cape May Branch from Newfield to Millville, a distance of about 10 miles. Near Westville, the railroad built its power plant. Three steam-driven turbogenerators individually rated at 2,000 kilowatts produced 6,600-volt, three-phase alternating current. Space was available for a fourth unit, if and when traffic demands required it. After being stepped up to 33,000 volts, the current was dispatched to eight substations located at regular intervals along the track. Rotary converters at each of these substations changed the alternating current to 650-volt direct current, at which voltage the current was fed into the third rail. With its 65 route-miles and 140 track-miles electrified, the West Jersey and Seashore surpassed the Long Island as the nation's longest steam railroad electrification.[7]

Since passenger operations alone were to be converted to electric traction, and no through cars were to be handled, the PRR chose to use multiple-unit cars rather than locomotives. It divided an order for 68 m.u. cars among three well-established firms: J. G. Brill of Philadelphia, American Car and Foundry (successor to Jackson and Sharp) of Wilmington, and Wason Car Manufacturing Company of Springfield, Massachusetts. The railroad itself developed the standard design upon which all the cars were patterned. Each unit measured 55 feet overall. Sixty-two of the cars were equipped as coaches and could seat 58 passengers. The remaining 6 were outfitted as baggage-mail combines. General Electric supplied the electrical apparatus, which included two 200-horsepower nose-suspended motors per car. The Pennsylvania specified that all the m.u.'s be built of wood—a curiously regressive stipulation in view of the all-steel rolling stock which it had ordered for use in the New York tunnels and for use by the Long Island Rail Road. Whereas the railroad spared no expense on the New York terminal project, it took every opportunity to avoid unnecessary expenditures on the WJ&S.[8]

Despite its parsimonious attitude, the PRR spent slightly more

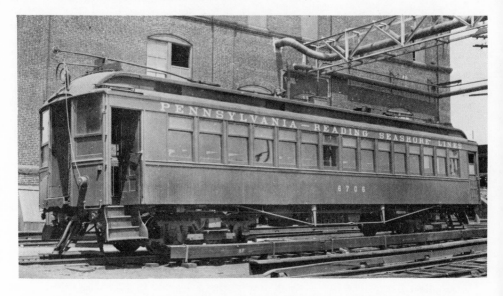

Wooden m.u. No. 6706 of the Pennsylvania-Reading Seashore Lines, successor of the West Jersey and Seashore, at Camden, New Jersey, 1947. (Harold
K. Vollrath Collection)

than $8 million to electrify its subsidiary. This figure encompassed
all the actual electrification work plus the construction of new terminals at Camden and Atlantic City, upgrading the roadbed, and installing a new automatic block signal system. On September 18,
1906, the first electrically powered train ran from Camden to Atlantic City. Soon m.u.'s were arriving and departing Camden on half-
hourly schedules. Fares dropped by an average of 20%, the price of
a Camden-Atlantic City round trip now being just $1.70. By 1911,
the West Jersey and Seashore was doing a thriving business the year
round. The railroad had to increase its m.u. roster to 93 cars to accommodate the growth in traffic, with 15 more cars (still mostly of
wood construction) on order.[9]

Most welcome in the Broad Street offices of the parent road were
the comparative cost computations for electric and steam operation.
The Seashore paid 18.19 cents per car-mile to operate and maintain
its m.u. fleet in 1910. During the final year of steam service over the
same route, the road had been spending 22.30 cents per car-mile.
Electrification had thus netted the WJ&S a savings of 4.11 cents per
car-mile so far. The cumulative savings for 1910, when the m.u.'s
ran some 4,552,532 miles, was $187,109. The economies wrought by
electric traction carried over to the railroad's balance sheet. Net
earnings for the WJ&S had not amounted to more than a few
hundred thousand dollars annually during the early years of the

century. By 1911, the road was regularly posting a yearly net income of about $700,000. The PRR attributed this upward trend primarily to electrification.[10]

Unfortunately, the Pennsylvania's conservative insistence on utilizing what even then was a technologically obsolescent electrification seriously undermined the value of the West Jersey and Seashore as an archetype for electric traction elsewhere on the system. At the very time the Pennsylvania considered the switch from steam to electricity on the Seashore, it was also debating the practicality of converting one or more of its main-line divisions to electric traction. These deliberations centered mainly—but not exclusively—on the Philadelphia, the Middle, and the Pittsburgh divisions. The tracks of the Philadelphia division ran westward from their namesake city through Paoli, Parkesburg, and Lancaster to Harrisburg. This line more or less traced the original PRR right-of-way of the pre-Civil War era. The Middle division stretched 130 miles between Harrisburg and Altoona. It generally paralleled the meandering course of the Juniata River. The 115-mile Pittsburgh division connected Altoona and Pittsburgh. The 35-mile segment of this division that lay between Altoona and Conemaugh scaled the summit of the Alleghenies and included the sharpest gradients on the PRR's east-west main line.

The turn of the century found the railroad enjoying a continuous upsurge of traffic between Philadelphia and Pittsburgh. Gross tonnage on this route had risen by 40% and ton-miles by 32% between 1899 and 1904. Passenger-miles ballooned in a similar manner. Business seemed almost too good. In 1903, President Cassatt informed his company's stockholders that "it is clear that on some of your lines, the capacity of the running tracks and yards has been reached. This is particularly true of your mainline between Pittsburgh and Philadelphia, where traffic has become exceptionally dense."[11] Cassatt and the company's directors subsequently decided to implement a three-year, $67 million improvement program aimed at spreading the flow of traffic between these two cities.

Items given the highest priority included four-tracking the entire length of the Middle division (except for a short stretch near the Spruce Creek tunnels) and enlarging the yards at Altoona and nearby Hollidaysburg. Since the addition of more tracks to the Philadelphia division main line was not feasible, the PRR constructed a new rail line east from Harrisburg. Over this—the Atglen and Susquehanna Branch, or the Low Grade Line, as it is more generally known—the railroad diverted most of its heavy freight trains. The western terminus of this line was at Enola, across the Susquehanna River from Harrisburg. There the railroad built a huge new classification yard. Freight trains leaving Enola ran south along the river for about 20 miles to a point opposite Columbia. There they veered east and, after having crossed the Susquehanna, passed south of Lancaster to rejoin the main line at Parkesburg. A short distance east of Parkesburg, near Thorndale, freight trains bound for the north Jersey or New York terminals left the main line again and switched onto the Trenton Cut-Off, following that line as far as Morrisville, where a junction with the New York division main line occurred. The PRR had built the Cut-Off to permit east-west through freights to bypass the congestion of the Philadelphia terminal area. Freight trains heading to or from Philadelphia and points south continued to use the main line from Parkesburg through Paoli to West Philadelphia.[12]

Further west the railroad attempted to alleviate another overcrowded situation with a similar freight detour. The primary obstacle to traffic between Harrisburg and Pittsburgh loomed a few miles west of Altoona, where the main line snaked around the Horseshoe Curve and battled a 1.86% grade in its assault on the eastern face of the Alleghenies. The four-track line there was continually clogged with trains making their way up or down the mountain. To break this bottleneck, the Cassatt administration ordered the double-tracking of the 18-mile New Portage Branch from the summit at Gallitzin down the mountainside to Hollidaysburg. At that point, trains could be routed to and from Altoona via a short connecting branch, or they could bypass Altoona altogether by following the Petersburg Branch, which connected with the main line some 40 miles to the east. In either case, the New Portage Branch would serve as an alternative to the Horseshoe Curve route. But even with this new line available to absorb excess traffic, a tremendous number of trains continued to traverse the east slope of the Alleghenies by way of the Curve. A survey conducted by the PRR during the last quarter of 1904, for example, revealed a daily average of 168 independent train movements around the Curve in each direction, including 28 passenger trains each way and numerous light helper movements.[13]

Officials in the road's operating department began to voice doubt as to the ability of both mountain routes to handle further increases in traffic. If business continued to grow at the same rate as it had over the past decade or so, the Pennsylvania's main stem across the mountains could well be taxed beyond endurance. The entire east-west flow of traffic could conceivably grind to a halt in a chaotic mass of jammed yards and motionless trains. The increase in ton-miles slackened a little over the next few years, however, allowing the railroad a brief respite during which it might consider various schemes for expediting its trains through the Allegheny barrier.[14]

One possible solution was to construct a third multiple-track line between Conemaugh and Altoona. Accordingly, the PRR dispatched surveying teams to locate the best passage for this new route. To no one's surprise, the surveyors indicated that the Pennsylvania was already using the two best rights-of-way, at least on the east slope. Therefore, construction of any new line having less severe grades than the current ones would necessitate the expenditure of enormous sums of money.[15]

Another solution might be to utilize electric traction in place of steam locomotives. Electrification offered several tempting advantages. Even the most ardent defenders of the steam locomotive admitted that electric locomotives could do without the costly army of shopmen and roundhouse gangs that were continuously swarming over steam power. No ashes need be dumped, no fires banked, no flues cleaned, no coal or water taken. Electrical equipment, being simpler in design and containing far fewer moving parts than reciprocating steam engines, was more reliable and easier to maintain. The experience of interurban railways had already proven this.

The electric locomotive's biggest attraction, however, was its pulling power. The electric motor had a much more favorable weight-to-horsepower ratio than did the steam engine. Thus an electric locomotive could produce more horsepower than could a steam locomotive of comparable weight. Moreover, on a short-term basis, an electric locomotive could nearly double its normal, continuously rated horsepower output by having its traction motors overloaded. These motors were so designed that they could deliver approximately 75% more horsepower than their continuous rating for short periods of time (generally no more than 1 hour). Only the amount of added heat which the traction motors could withstand restricted the degree and duration of the overload. Through the contact wire or the third rail, the locomotive could draw upon a virtually limitless supply of power.[16] For example, an electric locomotive normally rated at 3,000 horsepower could produce 5,000 horsepower or more for brief periods, such as when starting a train or ascending a grade.

The steam locomotive's horsepower, by contrast, was determined by the amount of steam which could be forced into its cylinders. Any quantity of steam produced in the boiler in excess of the cylinders' ability to use it was wasted. A steam engine, in other words, was physically incapable of carrying an overload of any kind. A greater horsepower rating did not automatically translate into superior drawbar pull, or tractive force; but here again the steam locomotive was at a disadvantage. As the speed of the locomotive increased, the pull at the drawbar decreased because the valve cutoff had to be shortened in order to stay within the steam generating capacity of the boiler. The tractive force of an electric locomotive also fell as speed increased. However, since horsepower is a product of tractive force and speed, and given the higher horsepower inherent in an electric locomotive, the maximum tractive force exerted by an electric was two or more times greater than the force exerted by a steam locomotive of comparable weight and speed (up to about 30 miles per hour).[17]

What these technical advantages meant, in sum, was that electric locomotives could haul longer, heavier trains than their steam counterparts, and at considerably higher speeds. A general rule of thumb (formulated by the electrical manufacturers) stated that an electric locomotive in freight service could pull a train weighing from 5 times the locomotive's own weight (on a 2.0% grade) to 16 times its own weight (on a 0.5% grade). This contrasted with the steam locomotive's ability to pull a freight train from 2 to 7 times its own weight on the same gradients. A similar ratio existed for passenger engines.[18] Therefore, converting any portion of the main line to electric traction would significantly expand its capacity while rendering more running tracks unnecessary.

In 1908, when PRR motive power men began to examine the possibility of introducing main-line electrified service, the advantages of using electric locomotives for heavy-duty, long-haul operations remained largely theoretical. The railroad, nonetheless, had enough faith in this new technology to conduct extensive studies of the use of electric traction both east and west of Harrisburg. Surveys were first made along the Low Grade Line between Harrisburg and Morrisville, a distance of 132 miles. The Pennsylvania calculated that traffic and maintenance requirements for this route necessitated the services of 171 Class H6b 2–8–0's, one of the road's standard freight types. On the other hand, studies showed that if the PRR were to electrify the Low Grade Line, only 50 electric locomotives would be needed. These estimates were based on the use of 150-ton motive power units similar to the New York Central's newly built S-class machines. Electrification would cost about $10 million but would save the railroad $750,000 annually in operating costs alone as compared with steam. If it switched to electric traction, the PRR could raise tonnage ratings by 30%; hence it would run fewer trains and save even more money.[19]

The Pennsylvania conducted a similar survey of its more mountainous Pittsburgh division. Again comparing an H6b with the rough equivalent of the Central's Class S engine, the railroad reckoned that there, too, maximum allowable weight of electrically powered freight trains could be increased by about one-third over those which were steam drawn. The heaviest westbound freights were then plodding up the east slope and around the Curve at no more than 10 miles per hour. A single electric locomotive could handle the same train at 30 miles per hour. Electrification of the Pittsburgh division would

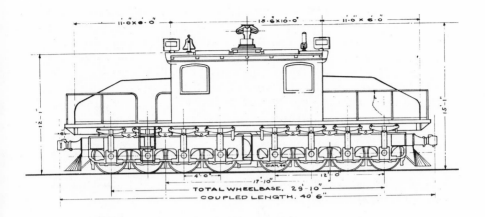

FREIGHT LOCO.–LOW GRADE FREIGHT LINE

WT. OF ENG. IN WORKING ORDER LBS.	300,000
TOTAL WT. ON DRIVERS LBS.	300,000
MAX. WT. ON ANY PAIR OF DRIVERS LBS.	37,500
METHOD OF DRIVE	GEARLESS
TRACTIVE FORCE LBS.	69,300

Drawing of the electric locomotive the PRR proposed to use on its freight lines east of Harrisburg before World War I. (PRR)

reduce operating expenses by $1.5 million per year. Fewer and faster freights would result in even greater savings. In reviewing the conclusions of the two studies, a committee of PRR engineering personnel headed by electrical engineer B. F. Wood noted that "ultimately our through freight lines must be electrically operated, especially those over heavy mountain grades, if we are to compete with the railroads having the advantages of lower grades."[20]

If at first glance the case for immediate electrification seemed overwhelming, several serious drawbacks existed which cast doubt on its feasibility. First there was the curse of the novelty of electric traction. Only a handful of American railways as yet had much experience with it. Fewer still could boast of having experience with electric traction in heavy-duty freight service. Only the Baltimore and Ohio, the New Haven, and the Great Northern (which in 1909 installed a 4-mile, 6,000-volt a.c. system through its new Cascade Tunnel) operated a significant number of freight trains electrically, and none of these had completed a truly long-distance electrification.[21]

Second, the installation and operating costs upon which the Pennsylvania formulated its projections were those of its own West Jersey and Seashore subsidiary and the New York Central. Both of these roads utilized low-voltage direct current. If the PRR were to electrify a whole division, in all likelihood it would elect to use alternating current, for which reliable data were not yet available.

Whether the railroad selected alternating or direct current, it could not elude the extremely high initial cost of electrification. The Wood committee had recommended that the Pennsylvania build its own power plants of sufficient size to supply not only railroad demands but those of on-line industries as well. It also recommended—perhaps with the WJ&S experience in mind—that certain money-losing branches which fed into the main line be electrified as well, with the expectation that electric traction would cut their losses. Both of these measures were designed to cushion the financial impact of wide-scale electrification.[22]

If the PRR did decide to electrify an extended portion of the main line, it would simultaneously be placing itself on the cutting edge of new railway technology. The road's board of directors undoubtedly realized that technological innovation does not inherently lead to economic gain for the innovator. The Pennsylvania's management could not guarantee that electrification's theoretical savings would appear as actual dollars and cents in the accounting department's ledgers. To put it another way, in the plain words of Andrew Carnegie, one of America's preeminent industrialists and one-time PRR division superintendent, "Pioneering don't pay."

The Pennsylvania's directors apparently gave more credence to the crusty Scot's maxim than to the optimistic forecasts of outspoken electric traction advocates like Frank Sprague and George Westinghouse. By 1913, the railroad was no longer expressing much interest in erecting catenary across the Low Grade Line or the Pittsburgh division. Instead, it was confining its inquiries to the possible electrification of the "Big Hill," that is, the 35-mile climb up and over the Alleghenies between Altoona and Conemaugh.[23] The distance was short enough so that if it so desired the PRR could minimize the "pioneering" aspects of the project by using low-voltage direct current. Also, the route-mileage was sufficiently low to permit the installation to be accomplished without a prohibitively high first cost.

Still the railroad hesitated. At the annual stockholders' meeting early in 1916, Samuel Rea—who had succeeded James McCrea as president two years earlier—explained this inaction.

> Electric traction would facilitate the heavy traffic movement on this difficult section of your Main Line and effect a saving in operating expenses, but the Company prefers to obtain the benefit of the experience of other lines in the use of electric traction for heavy freight trains, and to see a further expansion of its own revenues before procuring the new capital required for this important project.[24]

Revenues had been unsteady over the last several years. The railroad had posted its highest net income to date in 1912, when it took in slightly more than $42 million. Income had tumbled during the next couple of years, however. For 1914, the PRR netted only $34 million. It earned a return on investment that year of just 3.62%. And for the past half-dozen years, the road had been struggling in vain to reduce an operating ratio that stubbornly refused to drop below 80%.[25] In part, the Pennsylvania blamed its troubles on a business recession. A more persistent source of difficulties, it contended, was the Interstate Commerce Commission, which had refused to grant the PRR (or the railway industry in general) a significant rate increase since the Hepburn Act of 1905 had given it broader rate-making powers. On the other hand, the cost of doing business had risen markedly since that time. The result produced a situation that was not especially conducive to luring the investment capital needed for a large electrification project, and so the railroad bided its time.

This conservative stance persisted at the Broad Street headquarters into 1917. In 1916, revenues had increased somewhat, but not enough to convince management that the time was right for a sizable capital outlay. In fact, the rapidly expanding war in Europe had only aggravated the company's uncertainty concerning future revenues. President Rea told the stockholders early in 1917 that the war-related "scarcity of labor and the high cost of construction materials make it desirable not to urge this improvement." He pointed out, too, that the PRR was still in the process of "obtaining the experience of other lines in the use of electric traction for heavy freight and passenger trains."[26]

In spite of its cautious outlook, the Pennsylvania did take mountain electrification seriously enough to build a single prototype locomotive for such service. Working in close cooperation with Westinghouse, the railroad in 1917 outshopped a 1–C+C–1 side-rodded behemoth. While the PRR officially designated the monster Class FF1 and gave it road number 3931, railroad employees christened the new machine "Big Liz." "Big" was something of an understatement for this 250-ton giant, which was rated at 4,000 continuous horsepower and earned for itself the title of the "world's most powerful electric locomotive." The FF1, in a departure from traditional single-phase practice, employed a rotating phase converter to split the single-phase alternating current collected from the contact wire into three phases for use in the traction motors. The design of three-phase motors confined their operation to constant and uniform speeds, in this case, 10.4 miles per hour and 20.8 miles per hour. Constant speed operation presented numerous handicaps under most circumstances; but it was considered ideal at that time for work on mountainous grades, where its increased ability to lug tonnage counted for just about everything. A pair of FF1's, for example, could handle a 3,900-ton train westbound over the Alleghenies at 20.8 miles per hour, or a 6,300-ton train eastbound at the same speed. (The eastbound grade leaving Conemaugh was not so severe as the grade westbound from Altoona.) Another feature of the FF1's three-phase equipment that ideally suited it to mountain operation was regenerative braking. This characteristic permitted the locomotive's motors to act as generators when descending grades, thus producing rather than consuming electric current. (The surplus current was returned to the overhead lines to be used elsewhere.) The retarding effect of regenerative braking also provided a smoother, more efficient means than traditional air brakes of controlling the movement of a train downgrade and resulted in considerably less wear on brake shoes and wheels.[27]

The new FF1 quickly captured the fancy of the industry and of the general public. The trade periodical *Railway Review* affirmed the machine to be "the consummation of the best knowledge and experience that could be brought to bear on a problem of this nature." The *New York Times* described the engine as being as powerful as "a string of trolley cars a half-mile long," and patriotically proclaimed to its readers, "Once more has American genius demonstrated its supremacy in the railroad field!"[28] The FF1 may have been the product of American engineering genius, but it hardly represented a breakthrough in electric railway technology. In its use of side rods, split-phase electrical equipment, regenerative braking, and certain internal gear, it was merely a slightly larger version of the 3,200

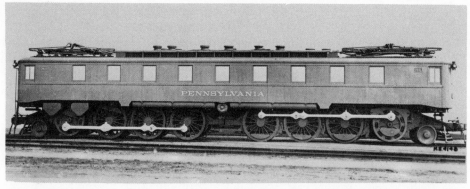

Builder photograph of the FF1 with pantographs down, 1917. (Altoona Public Library

horsepower LC-1 class freight locomotives that Westinghouse and Baldwin had built for the Norfolk and Western Railway. The coal-hauling N&W had experienced many of the same problems on its 27-mile Elkhorn (West Virginia) grade that confounded the PRR west of Altoona. Concluding that steam locomotives offered no solution to these difficulties, the railroad ordered a dozen LC-1's in 1914, all of which performed heroically from the very first day they attacked the fierce 2.0% grades of the Appalachians.[29] The Norfolk and Western had retained Gibbs and Hill as its electrical consultants. This fact, coupled with the Westinghouse company's role as equipment supplier and adviser to both roads, probably reinforced the conservative tendencies which the PRR had so far displayed in regard to electric traction. The Pennsylvania had followed the lead of the New York Central and several other railroads in entering New York City via low-voltage direct current. It patterned its single-phase installation at Philadelphia after the successful application of the same basic system on the New Haven. Now the PRR appeared ready to imitate the heavy-duty mountain electrification of the N&W.

The Pennsylvania even toyed with the notion of sending the FF1 to West Virginia for trials on the Elkhorn grade. After further deliberation, however, it decided to keep the locomotive close to home —in fact, almost within sight of the road's general offices. On August 24, 1917, No. 3931 took up duties as a helper engine on heavy freight trains ascending the 1.0% westbound grade between Philadelphia and Paoli. In this capacity, PRR and Westinghouse engineers would have ample opportunity to gather data which would be helpful in the event wires someday went up between Altoona and Conemaugh.[30]

Unfortunately for "Liz" and proponents of electrification, a number of factors intervened to preclude any possibility of erecting catenary over the mountains in the near future. First came the out-shopping in late 1916 of the prototype of the PRR's famed Class I1s 2–10–0 steamer. The Pennsylvania had always considered an improved steam locomotive to be a third option in the battle to break up the traffic jam on the Alleghenies. Initially, it had tried Mallets, ordering a 2–8–8–2 from the American Locomotive Company (Alco) in 1911 and an 0–8–8–0 from Baldwin a year later. Both engines worked out of Altoona as freight helpers. Neither one much impressed PRR motive power officials, and the Mallets were not duplicated. The railroad returned to two-cylinder power with the L1s 2–8–2, the first of that class rolling out from the Juniata erecting shops in 1914. This type, unlike the Mallets, was mechanically sound; but it was still not powerful enough to speed up the flow of traffic over the Alleghenies. The railroad was forced to continue double-heading its trains on either side of the mountain. Often a freight arriving in Altoona from the east had to be broken into two or more sections before it could be dispatched westward up the hill. Finally came the I1s in December 1916. The Pennsylvania was searching for a locomotive which could haul more tonnage than the L1s, yet use less steam and coal. In the new 2–10–0, the PRR had a locomotive that did just that. The I1s on the average exerted 30,000 pounds more tractive force than did the L1s, while consuming 12% less steam. The 2–10–0 was a triumph of PRR engineering acumen. By 1924, the railroad had a total of 598 of these engines on its roster.[31]

The year 1917 marked the beginning of mass production of the K4s 4–6–2, a class which brought the same kind of improvements to passenger service that the I1s brought to freight.[32] The success of these two new types of steam locomotives correspondingly lessened the urgency of electrification. America's entry into the European war in 1917 also worked against electrification. Under the dual burdens of swollen wartime traffic and the supervision of the federal government, the Pennsylvania had to halt experimentation with electric traction. The war might have proved a blessing to electrification. William Gibbs McAdoo, head of the new United States Railway Administration (USRA) which Congress created in 1918 to operate all American railroads, claimed that when the USRA took over the PRR "the congestion on the Pennsylvania lines . . . had just about brought freight traffic on that railroad to a standstill."[33] He did not mention that the underlying cause of this congestion was essentially not of the PRR's own making. Practically all the nation's war-related cargo was headed for northeast ports for transshipment to Europe. That an eventual blockage would occur should have come as no surprise, since the bulk of this traffic continued to follow the normal peacetime routes, that is, most of it was still routed via the Pennsylvania rather than on competing lines. The conditions thus created offered the perfect opportunity for electric traction to show what it could do in the way of expediting the movement of trains and reducing costs. Although McAdoo's agency pumped over $160 million in federal funds (more than 20% of all USRA wartime expenditures) into the beleaguered Pennsylvania, not one penny went for electrification. Clearly, the government was interested in immediate, short-term solutions only.

As the flood tide of freight and passengers began to ebb in 1919, it left in its wake a badly battered physical plant. The constant parade of wartime traffic ran roughshod over right-of-way and rolling stock, which suffered further from war-enforced deferred maintenance. To make matters worse, postwar earnings fell off sharply. In 1920, the railroad sustained a deficit in net railway operating income of over $62 million. Net income from all sources was only $32.8 million that year, the lowest since 1908.[34] The expenditure of large sums for electrification was out of the question. But after 1921, when the railroad reported a net income of only $24.3 million, business increased and the road once more began to feel the need to rid itself of the Allegheny obstacle course.

The PRR motive power department realized that the performance of the new I1s and K4s engines, however creditable, still left something to be desired. These locomotives were more powerful than their predecessors, to be sure; but since 1907, the Pennsylvania had been constructing all of its passenger equipment and an increasing percentage of its freight cars of steel. The heavier train weights which resulted canceled out much of the increased power of the 2-10-0's and 4-6-2's. Most passenger trains, for example, were routinely assigned two K4s locomotives westbound from Altoona. The heaviest limiteds often qualified for triple-headed power, at least as far as the summit at Gallitzin. Similarly, it was not at all unusual for a freight train to wind its way up the east slope powered by three or four thundering I1s's. After a brief flirtation with yet another Mallet design (a home-built 2-8-8-0, Class HC-1),[35] the railroad was fast coming to the conclusion that the only way it could avoid the excessive and expensive use of helper power over the Alleghenies was to electrify.

By the fall of 1923, rumors were again circulating to the effect that the PRR had determined once and for all to string wire on its Altoona-Conemaugh segment of main line. On November 9, a statement was released to the press from the Broad Street offices which seemed to validate this speculation. It read in part: "While no final plans have been developed for the running of trains by electricity on the mainline from the Conemaugh section to Altoona, it has reached the point of a definite intention to electrify the Allegheny Mountain line on both sides of the watershed."[36] In the same statement, the PRR went so far as to say that the generating stations for this electrification would be railroad owned, and that it was looking forward to negotiating long-term contracts with coal operators in the area.

Never before or since did electrification over the Alleghenies come so close to reaching fruition for the Pennsylvania. Once more, however, the scheme failed to materialize. For one thing, the railroad had not yet developed a satisfactory type of motive power for the job. It had discovered that the split-phase, constant speed characteristics of the FF1 were more trouble than they were worth. In 1924, the PRR outshopped a trio of Class L5 electric locomotives. These 200-ton machines carried a rating in excess of 3,000 horsepower and sported a 2-D-2 wheel arrangement with geared side-rod drive. Ostensibly built as successors to the DD1's in New York tunnel

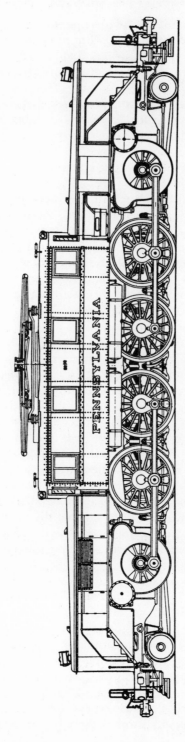

Side Elevation of One of the Locomotives. The Principal Dimensions are: Dia. of Drives 80 in.; Driving Wheelbase 22 ft. 3 in.; Dia. of Guiding Wheels 36 in.; Total Wheel-Base 54 ft. 11 in.; Length Inside Knuckles 68 ft. 2½ in.

Line drawing of the alternating current version of the L5. (*Railway Age*)

work, the L5's were also meant to succeed the FF1 as test-beds for mountain electrification. The railroad put into service one of the L5's—No. 3930, the only one equipped for a.c. operation—as a companion to the FF1 in Overbrook-Paoli freight helper service. The two d.c.-powered L5's went to New York to join the DD1's.[37] These locomotives required an extensive and lengthy period of testing before the Pennsylvania could be assured of their adequacy for rugged mountain operations. Electrification plans had to be delayed accordingly.

Representatives from Westinghouse (which had furnished the electrical gear for all PRR electric locomotives thus far, including the L5's) must have reminded the Pennsylvania at this time that the Norfolk and Western had been obtaining satisfactory results from its big LC-1's for nearly a decade under topographical and operating conditions that were remarkably similar to those confronting the PRR. So pleased had the N&W been by the results of electrification that by 1924 the road had nearly doubled the size of its original installation (to 55 route-miles) and purchased four giant (4,750 horsepower) locomotives (Class LC-2) to augment its fleet of a dozen LC-1's. Nor had any qualms about the proper type of motive power stopped the coal-hauling Virginian Railway from electrifying in 1924 its 134-mile Mullens, West Virginia-Roanoke, Virginia, main line, using single-phase a.c. power and Westinghouse electrical equipment.[38] But it was not within the tradition of the Pennsylvania Railroad to rely on the locomotive builders to dictate what kind of engine should be used under what conditions. Its motive power men had been custom designing their company's locomotives for nearly 75 years. They were not about to relinquish this practice just because they were dealing with electricity instead of steam.

In contrast to most of Europe, where electrification of steam railways was becoming commonplace, electric traction in the United States had made little headway by the mid-1920s. Other than the Virginian and the Norfolk and Western ventures, the only heavy-duty electrification of any distance was the 663-mile installation of the Chicago, Milwaukee, and St. Paul Railway through the Rocky Mountains and Pacific Northwest. In contrast to the eastern lines, the Milwaukee utilized high-voltage direct current, a system pioneered by General Electric that mirrored that firm's continuing faith in the superiority of direct current. Undoubtedly the so-called battle of the systems handicapped the efforts of the electrical manufacturers to win acceptance for this new form of motive power. At a meeting of the Philadelphia section of the American Institute of Electrical Engineers in 1912, for instance, the Pennsylvania's own B. F. Wood accused the a.c.-versus-d.c. controversy of being more responsible than any other factor for retarding the expansion of electrical railroading. Second to this dispute, Wood ranked the conservative (that is, pro-steam) attitude of most railway men. On this last point, many leaders in the electrical supply and commercial power industries agreed. The heated debates between a.c. and d.c. advocates paled in comparison to the bitterness and recrimination aroused when pro-steam and pro-electric forces clashed.[39] Few electric traction supporters claimed omnipotency for electric power, however. Most went out of their way to assure their opponents that the steam locomotive would always have a place on American railroads. George Gibbs's approach was typical. He told a meeting of the American Society of Mechanical Engineers in 1919 that

> the steam locomotive need not fear extinction or ever serious curtailment of its usefulness by reason of any electrical plans at present formulated. This being the case, there is ample field for steam locomotive engineers to go ahead with their program for perfecting the steam locomotive, with an assurance that the improvements made will be needed and welcomed.[40]

Pennsylvania Railroad mechanical officers as a rule did not allow themselves to be drawn into verbal combat for or against electric traction. In their view, ample room existed on the system for both forms of motive power. These gentlemen nevertheless held definite opinions, as a joint meeting of the New York sections of the American Institute of Electrical Engineers and the American Society of Mechanical Engineers revealed. This particular gathering occurred on October 2, 1920, to serve as a forum for discussion of the steam-versus-electricity issue. In addition to representatives from GE and Westinghouse, officials from the New Haven and the Norfolk and Western lines vigorously championed the cause of electrification. Not one PRR man spoke on the behalf of electrification. Rather, Alfred Gibbs and William F. Kiesel, Jr. (PRR mechanical engineer) were two of the most vociferous defenders of the steam locomotive. Gibbs argued that the steam engines and their facilities that were replaced by electric motive power represented a huge wasted investment. How could any railroad be persuaded that electrification would save money when in reality so much money was being wasted? He further contended that if branch lines were not included in the original electrification, steam locomotives with all their trappings would have to be retained. This was a wasteful duplication of equipment. Yet if the branches were electrified, traffic on most of them was so sparse electrification would never pay for itself, let alone bring a return to the railroad. For his part, Kiesel disputed the coal conservation claims of the electrified roads. He flatly stated that as far as he was concerned, no hard evidence had been produced showing that a steam railroad reduced its coal consumption markedly after having converted to electric traction.[41]

The views expressed by these men should not be confused with the rigid, narrow-minded opinions held by so many adherents of steam. After all, in comparison with other steam roads, the PRR was a leader in the use of electric traction. On the other hand, there is no reason to believe that, given their bias toward steam in 1920, Gibbs and Kiesel and their associates in 1915 or 1910 were any less inclined to favor steam over electricity for long-haul purposes. The Pennsylvania's engineering staff would not embrace electric traction for the Allegheny Mountain trackage until all the possibilities of doing the job with steam power had been explored and found wanting. As of the mid-1920s, the PRR still had faith in the steam locomotive's potential to resolve satisfactorily the road's traffic problems.

At the managerial level, these technological disputes were of less consequence. If the road's experiences at Philadelphia and New York were any indication, the PRR's executives seemed to have had a clearer realization of the merits of electric traction than did many of the company's engineers. What worried men such as Samuel Rea was not the technical argument, but whether the railroad would remain sufficiently prosperous for a long enough period of time to make electrification a worthwhile financial undertaking.

A more important factor which undermined mountain electrification, however, was the Pennsylvania's sudden recognition that traction could be utilized more efficiently on another part of the system. The railroad had already partially electrified its service in the New York and Philadelphia areas. Now increasing suburban business was forcing the PRR to make plans for extending its catenary north and south from Philadelphia. The railroad was also considering electrifying the main line between Manhattan Transfer and New Brunswick, New Jersey. In view of this boom in suburban traffic plus the steady growth in freight and passenger traffic throughout the northeast corridor, the Pennsylvania's management concluded that using catenary to link Philadelphia and New York would bring about a greater return on investment than would stringing wires along 35 miles of track in the middle of the Allegheny wilderness. Plans for electrifying the Altoona-Conemaugh segment were therefore held in abeyance as the railroad shifted its attention to the eastern end of the system.

5

Foundations for the Future

The Pennsylvania's trackage through the Allegheny Mountains may have been the focal point of electrification plans, but it was by no means the only section of main line where insufficient track capacity hindered operations. The four-track, 90-mile New York division between New York City and Philadelphia also experienced severe congestion. At first, the PRR attempted to apply conventional remedies, such as increasing the number of tracks and improving steam locomotive design. By the mid-1920s, however, the railroad had discarded the usual methods of attack and had committed itself to combating the ever rising traffic levels with electricity. Contrary to its heretofore cautious approach to electric traction, the Pennsylvania by 1929 had embarked on what was to become one of the most famous and most heavily traveled railroad electrification projects in the world.

A. J. Cassatt in 1903 originally suggested the construction of a freight-only line between the New York area and Morrisville, on the west bank of the Delaware River opposite Trenton. He believed that the opening of the new Pennsylvania Station in New York and the enlargement of freight terminals in northern New Jersey would so multiply the number of trains using the main line through that state that the existing four-track line would never be able to accommodate them all. No action was taken on this proposal until 1911, when the board of directors authorized the purchase of right-of-way for a double-track, 40-mile "relief line" between Colonia, New Jersey, and Morrisville. From Colonia, where the relief line was to join the main line, to Newark, the PRR planned to have six tracks.[1] At Morrisville, the Trenton Cut-Off began, thereby lessening the burden on the main line for the remaining 30 miles to Philadelphia.

Three years later, all the land had been secured and construction had commenced on a stone arch bridge spanning the Delaware from Lalor Street in Trenton to Morrisville. Six-tracking the main line east of Colonia was progressing smoothly. The Pennsylvania had incorporated the entire enterprise in the name of the Pennsylvania and Newark Railroad. Late in 1916, the same "scarcity of labor and high price of construction materials" which forced the railroad to put a moratorium on plans to electrify the Allegheny grade brought a similar halt to work on the Pennsylvania and Newark.[2]

After the war, the Rea administration had to decide whether to resume construction of the relief line or find a reasonable alternative. The introduction of the new K4s steam locomotive to this territory did allow the railroad to expand the size of its passenger trains and to improve schedules somewhat, but at best the K4s represented a stopgap measure. With train weights continuing to increase, double-heading would soon become necessary in order to maintain speeds competitive with other roads. Double-heading already was a not uncommon practice for freight trains. The powerful I1s, designed primarily for drag service, did not perform efficiently in fast merchandise service. These locomotives were more urgently needed west of Enola, anyway. Yet a single H-class 2–8–0 lacked the brute force which heavy-duty freight haulage demanded of a locomotive. Even the L1s (which did not appear east of Enola in substantial numbers until 1924) was not entirely satisfactory.[3]

Although the Pennsylvania had completed one or two studies of New York-Philadelphia electrification prior to the outbreak of World War I, most of its electric traction ambitions had centered on the Pittsburgh division. Even after the war, with the fate of the Pennsylvania and Newark hanging in the balance, the PRR still concerned itself chiefly with stringing wire over the mountains. Ironically, it was the federal government—the same entity that had foreclosed on the Pennsylvania's traction plans in 1917—which first recommended in the postwar era that the PRR electrify its main line along the eastern seaboard.

During the war, a number of large consumers of electricity in the northeastern United States feared that the frenzy of war-related activity would result in a serious power shortage in that part of the country. Among the most prominent worriers was E. G. Buckland, president of the New Haven. Early in December 1918, Buckland proposed to Secretary of the Interior Franklin K. Lane that his department conduct a survey of all potential power sources and users in the northeast. Such an inventory would be helpful in planning for future growth in a way that would avoid a catastrophic power famine in case of another national emergency. While Lane was receptive to the idea, Congress had more important matters to deal with just then. Finally, in 1920, Congress appropriated $125,000 to the United States Geological Survey (an agency under the Interior Department's jurisdiction) to make the study.[4]

The Geological Survey farmed out most of the research work to a group of independent consulting engineers, all of whom were specialists in particular fields, including electrical generation and transmission, electric railways, and rural electrification. W. S. Murray, the New Haven's chief electrical engineer, acted as the group's coordinator. As a member of its advisory board sat Elisha Lee, a vice-president of the Pennsylvania Railroad.[5]

Late in 1921, the Geological Survey submitted to Secretary of the Interior Albert Fall its report, *A Superpower System for the Region Between Boston and Washington*. Fall in turn delivered it to President Harding. The study as a whole displayed remarkable vision. It first outlined a "superpower zone," a strip roughly 150 miles wide stretching from Boston to Washington. By 1930, the report predicted, an investment of a billion dollars by the utilities would be required to supply all the power demanded in this zone. (Present investment by the utilities stood at $400 million.) To make sure that the unit cost of this electricity would be as low as possible, the panel of engineers urged that there be an increase in the size and number of steam and especially hydroelectric plants, the establishment of an interlocking grid system of power transmission among all power companies so that electricity could be dispatched quickly wherever needed, and the conversion of the maximum possible number of coal-supplied industries to electric current. These findings rested on economies of scale—the larger the generating stations, the more power consumed, the lower the production cost and hence the price of a kilowatt-hour of electricity.[6]

Among the most likely industrial candidates for conversion from coal to electricity, noted the report, were the steam railways. Of the 36,000 track-miles within the superpower zone, approximately 19,000 could be operated at a profit with electric traction. The survey pegged the cost of electrification at $570 million, but explained that since the 13 railroads involved were already spending $150 million annually on improvements, electrification, if spread over three or four years, should not prove financially burdensome. By 1930, the report estimated, these railroads would be saving $81 million yearly over the cost of steam operations. Labor and maintenance costs would drop substantially, train speed and tonnage would increase, and coal consumption would be reduced by 10.5 million tons per year. The report singled out but one main line in the New York-Philadelphia corridor—that belonging to the PRR—for electrification. Traffic on the parallel B&O/RDG/CNJ main stem should be diverted to the Pennsylvania in order to obtain the maximum return on fixed-plant investment. Similar traffic diversions were recommended for other corridors, for example, Boston-Albany.[7]

What impact the superpower survey had on the PRR is difficult to assess, since neither the President nor Congress took action on the study's conclusions. The large capital investment required of the utilities and related industries discouraged hasty actions on their part. Furthermore, any move toward consolidation of traffic on key through routes might have raised serious challenges involving the cooperative relationships among traditionally competitive railways and the regulatory barriers which heretofore discouraged collective agreements between these companies. The survey did reveal some interesting statistical comparisons between PRR steam and electric motive power within the superpower zone. For 1919, the Pennsylvania spent 16.7 cents per locomotive-mile to operate its fleet of 33 DD1's, or $7,266 per locomotive per year. This figure included labor, mechanical maintenance, and power costs. Prorated for a locomotive weighing 100 tons on drivers (each DD1 had 103 tons on drivers), the PRR was spending 16.3 cents per locomotive-mile or $7,088 per locomotive per year. Operational costs for all PRR steam locomotives within the zone averaged 45.5 cents per locomotive-mile, or $10,850 per locomotive per year. Prorated for 100 tons on drivers (the steam locomotives had an average of 83.5 tons on drivers), the Pennsylvania was spending 53 cents per locomotive-mile or $11,152 per locomotive per year.[8] Granted, the railroad was still using many elderly 2–8–0's and 4–4–0's in the superpower region, but the bulk of its steam roster consisted of 2–8–0's and 4–4–2's of relatively recent (post-1910) vintage, plus brand new 4–6–2's. Nor could the DD1's be accused of holding light-duty assignments. Battling the uncompromising 1.93% tunnel grades and wheeling 14-car limiteds at mile-a-minute speeds compared favorably with the most rigorous chores handled by PRR steam engines anywhere in the zone.[9]

Whatever the influence of the government's survey, there could be no denying that after the war the Pennsylvania gradually came to regard its New York-Philadelphia corridor as ripe for conversion to electric motive power. Evidence that the Rea administration viewed electrification as more than just an expedient to overcome a mountain bottleneck appeared even before Congress had appropriated funds for an electric power study. In 1919, President Rea directed that all present mechanical officers be trained thoroughly in the fundamentals of electric traction; henceforth no mechanical officer would be permitted to achieve senior status until he had acquired a solid understanding of electric as well as stream traction.[10]

A more tangible indication of the Pennsylvania's growing inclination to apply electric motive power in the northeastern corridor came in 1924 with the construction of the three Class L5 locomotives. Because these units appeared at a time when the railroad began to waver in its resolution to electrify the Allegheny grade, their technical composition tended to reflect their owner's ambivalence toward the most suitable site for future electrification. James T.

L5 No. 3930 on the Philadelphia-Paoli line, 1924. (Altoona Public Library)

Wallis, PRR chief of motive power, and his staff designed the L5. They hoped to develop a "universal" locomotive which would be equally at home forwarding crack flyers through the Hudson tunnels, dragging coal trains around Horseshoe Curve, or wheeling merchandise expresses between New York and Philadelphia. By possessing such a general utility machine, the Pennsylvania would have a satisfactory motive power unit available no matter where it eventually chose to electrify.[11]

The railroad outfitted its first L5, No. 3930, with a.c. electrical gear. The locomotive's four Westinghouse Type 418 motors were the largest single-phase traction motors yet built in the United States. The drive mechanism was of the gear and side-rod type similar to that employed in the FF1. Each pair of motors was geared to a jackshaft, which in turn was connected via side rods to a pair of drivers on either side of the locomotive. The L5 thus had a total of eight driving wheels, plus a single nonpowered axle at either end, to give it a 1–D–1 wheel arrangement. The styling of the superstructure represented an interesting break with the traditional box cab designs of the FF1 and DD1. The L5 retained the angular lines of its predecessors but incorporated a rectangular steeple cab in the center of the frame (directly above the drivers) and lower, more elongated hoods at each end. The railroad installed d.c. equipment (again of Westinghouse manufacture) in the second and third L5's, Nos. 3928 and 3929, for experimental operation in the New York zone. The a.c.-powered unit entered trial service as a helper on the Paoli electrification. The question of current supply was of little importance, however. In line with the Pennsylvania's quest for universality, the locomotives could be adapted easily for alternating or direct current without having to undergo extensive structural modifications.[12]

The advent of the L5 was the first unequivocal indication that the PRR's management was in the process of shifting the focus of electrification from the Allegheny Mountains to the New Jersey corridor. Still, Chief of Motive Power Wallis and his associates did take into account possible Altoona-Conemaugh service when they designed the new engine. As the L5 was taking shape on the Altoona drawing boards in 1923, erection of catenary over the mountains seemed likely. Thus the L5's had to be exceptionally rugged and powerful machines.

Nevertheless, as the new locomotives rolled out from the Altoona shops in 1924, their ability to perform capably in mountain service was questionable. They were powerful, to be sure. The railroad's engineers estimated that two a.c.-powered L5's could haul a 4,750-ton train up the east slope at a steady 24 miles per hour. Even so, the locomotives did not come close to attaining the rating given a pair of FF1's (6,300 tons at 20.8 miles per hour) or any of the electrics then in service on the Norfolk and Western or planned for the Virginian. The L5's 80-inch driving wheels—the largest ever applied to a PRR electric locomotive[13]—combined with 750 horsepower per driving axle would have made the engines much too slippery for the low-speed, heavy-tonnage kind of operation which predominated on the line between Altoona and Conemaugh. The railroad could have installed special gearing to partially compensate for this deficiency, but only at the cost of the locomotives' versatility. Special gearing would have restricted the L5's sphere of operations to the Alleghenies just as assuredly as the constant-speed characteristics of FF1 limited that machine's usefulness. The large drivers and high axle horsepower were much more suited to a locomotive whose purpose was to pull fast passenger and freight trains across the relatively flat terrain of the New York division.

The same might be said of the L5's 22-foot, 3-inch wheelbase. Although the DD1 use an articulated frame, PRR engineers had not entirely abandoned their suspicion of articulation as a source of tracking instability. They had chosen not to apply it to the L5.[14] Yet such a long, rigid wheelbase could very possibly have caused tracking difficulties on the serpentine route over the mountains, as well as elsewhere on the tortuous Pittsburgh division or on the Middle division. The New York division, by contrast, contained few restrictive curves. These technical aspects of the L5, then, indicated that the demands of the New York-Philadelphia corridor weighed more heavily on the PRR's motive power planners than did the requirements for a short-haul mountain electrification.

The railroad officially claimed that the new d.c.-powered L5's were intended primarily to augment the DD1 roster in the New York zone. (This area, with its western terminus at Manhattan Transfer, had become a separate operating entity from the New York division. The New York division extended from Manhattan Transfer to Philadelphia's Mantua Junction.) Long-distance passenger business to and from New York was growing steadily during the 1920s. The addition of only two more locomotives did not perceptibly relieve the strain on the older DD1's. In a January 4, 1926, request for authorization to build eight more L5's, James Wallis warned Vice-President Elisha Lee that "the increase in business handled in New York is such that we do not have sufficient electric locomotives to take care of the shipping for class repairs and periodic inspection."[15] The Long Island Rail Road was also clamoring for some heavy-duty motive power. The Pennsylvania decided to provide its subsidiary with a few DD1's rather than with more expensive new units. This action only intensified the need for more engines for Hudson tunnel service.

Wallis hinted at the PRR's long-term intent for the L5's a few months later. In still another petition for even more locomotives (15 L5's costing $130,000 each), he reminded Lee that while these units would be equipped for d.c. operation, they could easily be converted to a.c., thus making them ideal for New York-Philadelphia through service. Westinghouse was more candid—and optimistic. In announcing a PRR order for electrical equipment for four more L5's, a Westinghouse spokesman declared that the order represented a preliminary step in the railroad's plan to electrify the entire main line from New York to Washington.[16]

Vice-President Lee and the railroad's directors shared Wallis's confidence, if not Westinghouse's bold assuredness. By the end of 1927, the railroad had in service or on order 24 L5's. Unit No. 3930 remained the lone a.c. machine. All 21 of the new L5's were equipped for d.c. operation, and all featured the distinctive center cabs. But unlike the original trio, the newer units were delivered with hoods nearly as high as the cab itself. This structural modification was yet another indication that the PRR anticipated using these locomotives in long-distance service. The a.c.-powered L5 initially had one pantograph mounted on its cab roof. In order to enhance the machine's reliability, however, PRR engineers decided to remove the single pantograph and position one on each hood. Since the low hoods were not of sufficient height to enable the pantographs to reach the contact wire overhead, No. 3930 had to be withdrawn to the shops and fitted with higher hoods. Therefore, although the later L5's might temporarily utilize direct current in New York area operations, the Pennsylvania constructed them with the higher hoods, too, with the expectation of ultimately converting them to alternating current for use in long-distance service.

The difference between the Class L5 engines and earlier PRR electrics was not confined to the center-cab styling. For the first time in the history of its heavy electric traction program, the Pennsylvania bought electrical gear from companies other than Westinghouse. General Electric provided the components for four locomotives. American Brown-Boveri, the North American subsidiary of the Swiss engineering firm of Brown-Boveri, outfitted seven more. The end of the sole reliance upon Westinghouse was another signal that electrification of the New York division was imminent. To complete such an ambitious project successfully, the railroad would have to call upon the technical expertise and manufacturing capacity of more than one supplier. In an effort to hold down costs, the PRR initially insisted that many of the electrical and mechanical parts of the L5's be interchangeable, regardless of manufacturer. This would have constituted an accomplishment in view of the highly individualized nature of the electric railway supply industry, had not certain technical difficulties forced the railroad to cancel its directive. The locomotives were subsequently equipped with similar, though not identical, components, an arrangement that was to cause problems that the Pennsylvania's engineers would not soon forget. The PRR itself assembled all carbodies and mechanical fittings at Altoona.[17]

The L5 did not in itself prompt the Pennsylvania to shift its electrification interests to the New York-Philadelphia line. The locomotive merely symbolized this change. Electric traction undoubtedly would go far toward relieving congestion on the New York division and bringing about increased efficiency of operation, but it would yield basically the same benefits if applied to the Altoona-Conemaugh territory. At least two factors influenced the railroad's eventual decision to foresake a limited mountain electrification in favor of a more extensive electrification further to the east.

The first factor related to the PRR's suburban service at Philadelphia. The electrification of its Paoli and Chestnut Hill lines was so successful that in 1924 it strung wires along a third route, the Fort Wasington Branch, which diverged from the Chestnut Hill line at Allen Lane and continued north for about 6 miles to White Marsh.[18] Steam engines remained the only type of motive power on three other key suburban routes: north via the New York division main line to Trenton, south via the Maryland division main line to Wilmington, and westward along the West Chester Branch. PRR planners realized that even though these three routes would one day be converted to electric traction, the antiquated facilities at Broad Street Station would still impede the efficient handling of local traffic. Consequently, in 1923, shortly after a fire had destroyed the mammoth Broad Street train shed, the Pennsylvania's directors began considering plans to replace the old terminal. The following year, PRR and Philadelphia municipal officials initiated negotiations regarding a joint railroad-city project which called for the construction of a new through station on the west bank of the Schuylkill at Thirtieth and Market streets and a new station solely for suburban traffic to be located downtown not far from the old Broad Street Station. The upper floors of the suburban station would also house the PRR's general offices. Broad Street Station was to be razed, along with the venerable "Chinese wall." In 1925, the railroad added a third structure to its proposed improvement program, a 14-story building at the corner of Thirty-second and Market streets which was to provide office space for engineering and other lower echelon personnel. The railroad had also finally determined to place the trackage at its suburban station below street level in order to use a minimum amount of expensive surface property. Because of the subterranean nature of this new terminal, the Pennsylvania concluded that the time had come to electrify the three remaining primary commuter arteries.[19]

In conjunction with the Hudson and Manhattan Railroad, the Pennsylvania was also planning to construct a spacious new terminal at Newark. The station was to replace Manhattan Transfer as the interchange point for steam and electric motive power and as the place where PRR riders made connections with trains traveling to and from lower Manhattan. In addition, ever since the Paoli electrification had proven itself, the railroad had been weighing the possibility of instituting similar m.u. service from New Brunswick at least as far as Manhattan Transfer, a span of 23 route-miles.[20] Assuming that the intensifying demands of suburban traffic would make electrification to New Brunswick inevitable in the not too distant future, would it not make sense to convert to alternating current the line between Manhattan Transfer and Pennsylvania Station in connection with the improvements at Newark? If this were done, only a 25-mile gap—the distance between New Brunswick and Trenton—would be left without catenary over the length of the New York division, as compared to about 80 route-miles remaining unelectrified on the Pittsburgh division in the event electric traction were applied on the mountain west of Altoona.

The second factor favoring New York-to-Philadelphia electrification pertained to the matter of power supply. The Pennsylvania intended to build and operate its own generating station to furnish power for electrification over the Alleghenies. On the other hand,

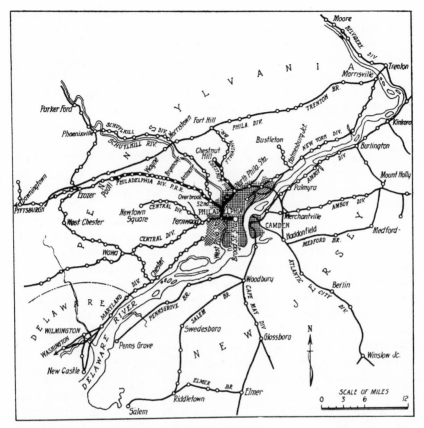

PRR Philadelphia suburban lines. *(Railway Review)*

the railroad already had a power plant (at Long Island City) which could provide current to the northern end of the New York division. On the southern end, the PRR had been purchasing large quantities of power from the Philadelphia Electric Company for a decade. To meet the expanding needs of residential and industrial customers in the post-World War I period, Philadelphia Electric was in the process of enlarging its generating capacity and welcomed an additional railroad load to help offset construction expenses and fixed costs of operation.[21]

In June 1924, Philadelphia Electric acquired an option to purchase the Susquehanna Power Company. That firm, merely a paper corporation controlled by the investment house of Bertron, Griscom and Company, owned some 5,400 acres of land in the vicinity of Conowingo, Maryland, a picturesque hamlet situated on the east bank of the Susquehanna River just south of the the Pennsylvania state line. Susquehanna Power also owned the hydroelectric rights along this stretch of river. Philadelphia Electric contemplated building a colossal $50 million generating plant there, a hydroelectric installation second in size only to that at Niagara Falls. Speculation was widespread in the utility industry that Philadelphia Electric's interest in Conowingo stemmed in large measure from the utility's desire to influence the Pennsylvania Railroad's electrification plans. The existence of an ample supply of electricity was crucial in determining the future of electric traction on the Pennsylvania. Conowingo's potential as a source of relatively cheap, plentiful power could well persuade the PRR to extend electrification in the Philadelphia region rather than further to the west. Philadelphia Electric officials refused to confirm or deny that this was their company's intention. The railroad professed ignorance about the whole Conowingo transaction.[22]

A few days after spokesmen for the Pennsylvania had disavowed any knowledge of the utility's dealings with Susquehanna Power, however, other PRR officials did admit that their company had a detailed study under way of New York-Washington electrification. Nevertheless, they insisted that electric traction was still most likely to be employed between Altoona and Conemaugh, not along the eastern seaboard. In January 1925, the railroad announced the completion of its investigation of extending electrification all the way to the nation's capital. It estimated the cost of equipping all 225 route-miles with single-phase a.c. catenary, purchasing new rolling stock, and upgrading the right-of-way to be about $50 million—too large a sum, the road said, for the project to be financially practicable at that time. The Pennsylvania's statement made no mention of the Allegheny mountain electrification, a significant omission.[23]

Meanwhile, the Philadelphia Electric Company exercised its option to purchase the assets of the Susquehanna Power Company. The utility estimated that when completed, the hydroelectric plant at Conowingo would produce over a billion kilowatt-hours annually. It represented a capital investment savings of $21 million over steam plants of equivalent capacity. The cost to generate each kilowatt-hour would be cheaper, too, since Conowingo would have no need of the 750,000 tons of coal per year that these steam plants would burn.[24]

Contractual relations between Philadelphia Electric and the Pennsylvania Railroad had thus far been mutually satisfactory. Throughout its 14-year association with the utility, the railroad had never expressed any discontent regarding Philadelphia Electric's rate structure. Indeed, in the spring of 1924 the utility had put into effect its third rate reduction (for all customers) in as many years.[25] The PRR had judged the electric company to be very ably managed from both financial and engineering standpoints. Utility officials had always been most cooperative in meeting the special requirements of the railroad. In return, the Pennsylvania was Philadelphia Electric's second largest customer (after the Philadelphia Rapid Transit Company) and consumed 15% of the utility's output.[26]

In view of this apparently harmonious relationship, even some of the most knowledgeable observers in the business community were stunned on February 10, 1927, when the PRR disclosed its intention to construct a massive power plant of its own to generate all the electricity it needed for a New York-Philadelphia electrification. Railroad officials claimed that they had met with executives of Philadelphia Electric, Philadelphia Rapid Transit, and the United Gas Improvement Company (UGI) to work out details for a $10 million coal-fired station to be located at Trenton to generate 25-cycle power exclusively. The PRR maintained that by itself, it could not afford such a venture, but if it could secure a cooperative agreement to finance the plant, there would be no real reason not to electrify the entire New York division.[27]

Philadelphia Electric president Walter H. Johnson quickly denied that his company had taken part in these negotiations. He said that he was shocked by the very notion that the PRR might generate its own power. Once the Conowingo station came on line, Philadelphia Electric would supply more than enough electricity for all its customers' requirements, and at low rates. Johnson could see "no advantage in [making] capital expenditures to obtain results which already existed." The transit company also disavowed being privy to the talks. Only Arthur Thompson, UGI's chief executive officer, admitted that his firm had been a party to conferences with the railroad—but he staunchly asserted that Philadelphia Electric and Philadelphia Rapid Transit had participated, too.[28]

Wall Street regarded these maneuverings as part of a corporate chess game. United Gas Improvement, a holding company, owned most of the suburban utilities which ringed the city of Philadelphia. It had been trying to add Philadelphia Electric to its organization since 1913. Its most recent proposal to absorb the firm came in 1926, but Philadelphia Electric's board of directors refused the offer. United Gas Improvement subsequently used the proposed Trenton generating station as leverage to force the electric company back to the bargaining table. President Johnson and his board apparently conceded under this pressure, for in May 1927, UGI and Philadelphia Electric representatives resumed negotiations aimed at consolidating the two companies. Four months later, the two sides had reached an agreement in principle. In February 1928, control of Philadelphia Electric passed to UGI to form the largest consolidation of utilities yet attempted in the United States. Philadelphia Electric retained its corporate identity but henceforth all major decisions were made in the UGI boardroom.[29]

The Pennsylvania Railroad claimed to be a neutral party in these dealings. It had no objections to a UGI-controlled Philadelphia Electric and in fact had representatives on the boards of both firms. Nonetheless, the PRR undoubtedly realized that it would probably benefit slightly from the amalgamation. Philadelphia Electric traditionally had been reluctant to expand its interests beyond the city proper, thereby denying itself certain economies of scale which would have resulted from a single utility serving not only the city but its environs as well. The takeover of Philadelphia Electric by UGI, combined with a similar absorption of several New Jersey-based utilities, formed a single integrated power system in the Delaware Valley which could well result in more favorable rates for the railroad. The consolidation met with favorable reaction from the general public, too, which respected both UGI and Philadelphia Electric as progressive, civic-minded enterprises.[30]

Events occurring during the next year or so reflected the gathering momentum of the PRR's electrification plans. The wrangling over the Trenton power plant ceased as suddenly as it had begun. The railroad and Philadelphia Electric in July 1927 signed a 20-year contract in which Philadelphia Electric pledged to supply all the electricity the Pennsylvania needed for present and future operations within the utility's territory. Philadelphia Electric in effect guaranteed to meet any load requirement set by the railroad—an unprecedented concession. President Walter Johnson was well aware of the significance of his company's pact with the PRR. "The signing of this contract," he stated, "the most important ever executed between a railroad and a public utility, means a new era in railroad electrification." Johnson was confident that Conowingo (expected to be in operation by 1928) plus the expansion of the utility's Richmond station (a coal-fired facility in Philadelphia) would be more than adequate to satisfy the railroad's demands.[31]

At about the same time the contract was made public, yet another feasibility study of New York-Washington electrification was laid before the PRR's executives. Like its predecessors, the report duly noted that electric traction would increase track capacity, lower maintenance costs, and simplify operations in general. This latest study, however, in particular stressed electrification's traffic-building potential. Electrification would pay for itself not only through reduced costs of operation (which heretofore had been the railroad's primary concern), but also by attracting more freight and passenger traffic. Since rates were fixed by law, improving service (by providing faster schedules, for example) was practically the only means

Conowingo hydroelectric plant. (Philadelphia Electric Co.)

available that a railroad could use to increase business and consequently revenues. Electrification, in other words, offered the PRR an opportunity to take an offensive stance to counter the rising challenge for intercity traffic from airplanes, trucks, buses, and private automobiles.[32]

The Pennsylvania's management was particularly alarmed by the competition offered by buses and automobiles. The astounding proliferation of these vehicles (especially autos) since World War I had already begun to have an adverse effect on the road's passenger revenues. Fortunately, commercial aviation still had to overcome a great many technological obstacles before the PRR felt the impact of the airplane on intercity traffic. The company's executives were also concerned about the truck's potential to lure away a sizable amount of freight business, but later events would show that in the 1920s they believed the most serious threat to be in the realm of passenger service. The Pennsylvania does not seem to have regarded competition from other railroads as a compelling reason for electrifying. The PRR and the New York Central had many years before tacitly resigned themselves to a virtual stalemate for the New York-Chicago passenger trade. Unless electrification were extended over a very long distance, it would have little effect on this standoff. And in the New York-Washington corridor, the Pennsylvania's long dominance over its lone rival, the B&O/Reading lines, showed not the slightest hint of weakening.

Even without a long-distance electrification, the Pennsylvania in many respects was doing very well, in spite of the advent of competitive forms of transportation. In 1921, at the nadir of the postwar recession, the railroad had launched a system-wide campaign to bring about greater efficiency of operations. Six years later, this quest was bearing fruit, at least in the area of freight traffic, as a comparison of a few key statistics reveals. By 1927, the PRR's operating ratio—the proportional relation of operating expenses to operating revenues —had declined for the seventh consecutive year (to 76.9%). Gross ton-miles per train-hour had climbed 26.4% in that same period, miles per freight car day were up 27.1%, and pounds of fuel consumed per 1,000 gross ton-miles decreased 11.7%%. The crusade for increased efficiency was made easier by the prosperity which the railroad and the nation were enjoying. Net railway operating income for 1927 had reached $110 million, nearly triple that of 1921. The PRR recorded a net income for 1927 of slightly over $68 million—the highest in the company's 79-year history.[33]

Much of the personal energy behind this efficiency drive came from William Wallace Atterbury, who had succeeded the retiring Samuel Rea as head of the PRR on October 1, 1925. Like many of his predecessors, Atterbury had worked his way up through the engineering ranks. He was born in New Albany, Indiana, in 1866, the seventh son of a lawyer turned Presbyterian home missionary. Twenty years later, upon graduation from Yale's Sheffield Scientific School, young Atterbury enrolled as an apprentice at the PRR's Altoona shops. He held a number of engineering posts in the years that followed and in 1901 became general superintendent of motive power with headquarters at Altoona. In this position he impressed President Cassatt so much that he advanced Atterbury over many senior executives to appoint him general manager of Lines East in 1903. By 1911, Atterbury had risen to fourth vice-president. When the practice of numbering vice-presidents was discontinued a short time later, he became "Vice-President in Charge of Operations."

In 1916, Atterbury was elected president of the American Railway Association (forerunner of today's Association of American Railroads). There he attracted the attention of the War Department, which was searching for someone to take charge of the French railroads as part of General John J. Pershing's American Expeditionary Force. In 1917, President Woodrow Wilson commissioned Atterbury a brigadier general and designated him "Director-General of Transportation" under Pershing. The PRR executive worked near-miracles with the war-ravaged rail network of the Allies, winning the Distinguished Service Medal and several foreign honors for his efforts. This was a phase of his career of which he was exceptionally proud. Even after he had assumed the presidency of the PRR, he was still often referred to as "General" Atterbury.

After resuming his domestic railroad duties in 1919, Atterbury continued to attract attention. He helped to implement the new regional system of organization in place of the outmoded and inefficient Lines East and West scheme. He led his railroad's successful struggle to rid itself of national shop-craft unions and in the process became the chief spokesman for what many observers termed the antilabor sentiments of all American railroads. In truth, Atterbury firmly believed in the right of workingmen to organize and bargain collectively, but, in the tradition of PRR management, he held that each railroad could best deal with its own employees rather than with representatives of an industry-wide union.[34]

Atterbury had long been convinced that if railroads were to offer the most efficient service possible, they had to become total transportation companies. Under his leadership, the PRR acquired controlling interests in bus lines, airlines, and trucking firms. Again in pursuit of efficiency, Atterbury was in the forefront of the movement during the late 1920s to consolidate American railroads into a comparatively small number of regional systems. He viewed electrification in a similar light. It was an important technological tool which the Pennsylvania could use to lower costs, improve service, and provide the ultimate in transportation efficiency. Even when compared with his illustrious predecessors, W. W. Atterbury was surely, as the *New York Times* called him, "a forceful and original figure in railroading."[35]

By 1928, then, the Pennsylvania Railroad had the prerequisites necessary to undertake an expensive electrification project. It had experienced, progressive leadership not only in Atterbury, but in his top three lieutenants, Vice-President Elisha Lee, Financial Vice-President A. J. County, and Vice-President in Charge of Operations Martin W. Clement. The railroad was financially sound and its future seemed bright, although there were certain signs that forewarned of difficulties. Passenger-miles, for example, declined 10% between 1921 and 1927. Gross passenger revenues had fallen to $140 million, down $15 million from 1921. Yet where could the economies of electric traction be better applied than in the New York-Washington corridor, which generated more passenger-miles than any other main line of comparable length in the country? While gross receipts for freight traffic showed a steady upward trend, average revenue per ton-mile just as steadily decreased, going from 1.180 cents in 1921 to 1.022 cents in 1927. Electrification, by cutting the cost of transporting freight, would proportionally boost the rail-road's net income on a ton-mile basis.[36]

Both the Rea and the Atterbury administrations had worked dili-gently to prepare their company for a major program of capital im-provements. The PRR had issued no bonds since 1921, and no bonds were scheduled to mature until the late 1930s. Most postwar improvements had been paid for from earnings. At the same time, the railroad had been steadily reducing its equipment trust obliga-tions and other portions of its funded debt so that it would not incur undue hardship if it had to borrow large sums to finance electrifica-tion.[37]

From a technical standpoint, too, the Pennsylvania was ready to assume the burden of a major electrification. The long-term contract with the Philadelphia Electric Company provided a secure base from which to draw power. An adequate supply of electricity for the PRR was further ensured in 1928 when Philadelphia Electric, Pennsyl-vania Power and Light, and Public Service Electric and Gas (New Jersey) agreed to interconnect their systems. The three utilities thus established one of the world's largest power pools, with a combined generating capacity of 2,250,000 kilowatts.[38] The 25-cycle, single-phase alternating current system which the PRR would utilize had become the standard design of heavy-duty American steam railroad electrification by the mid-1920s. The railroad no longer needed to fear the "pioneering" qualities of alternating current.

The only significant area in which the Pennsylvania would have to play the role of technological innovator was motive power. The rail-road still lacked a satisfactory locomotive for long-distance service. The L5 had not lived up to the railroad's expectations. Its long rigid wheelbase produced inferior tracking qualities even in limited New York zone operations. Even more serious were the troubles with the Type 418 traction motors. Westinghouse had developed these motors specifically for installation in the L5's; consequently, they had not been proven beforehand in everyday service in other locomo-tives. They suffered from numerous failures, high repair costs, and general unreliability. The inability of L5's of one manufacturer to accept parts supplied by another manufacturer further inflated the costs of repair and maintenance and resulted in far more downtime than the railroad was willing to accept. The jackshaft and side-rod arrangement with which the machines were fitted was likewise ex-pensive to maintain, besides being mechanically inefficient and un-suitable for prolonged high-speed running.[39]

The main obstruction to the development of a satisfactory type of motive power was the inability of the electrical manufacturers to produce an a.c. motor small enough to fit between the locomotive's driving wheels (thus eliminating the need for jackshafts and side rods) and yet large enough to produce sufficient horsepower for heavy-duty service at high speeds. The New Haven had sidestepped this problem by using many electric locomotives of only moderate horsepower and tractive force (EP-1's) in a "building block" concept. Multiples of these engines could be made up to meet varying train weight requirements. The PRR did not care to imitate this practice. Clinging to the orthodoxy of its own steam locomotive designs, the Pennsylvania still preferred to achieve a high horsepower rating using as few driving axles as feasible—even though this custom resulted in the flawed L5's. In 1928, the Pennsylvania directed Westinghouse and General Electric to concentrate their energies on developing a small but powerful a.c. motor that could meet the railroad's needs. At once, the two manufacturers independently began intense research and development programs.[40]

The Pennsylvania, confident that Westinghouse and GE would be successful in their efforts, decided to take the final, decisive step. On November 1, 1928, President Atterbury formally announced his company's intention to electrify the main line between New York and Wilmington, Delaware. The railroad also planned to extend catenary westward from Philadelphia along the Low Grade Line. Seven or eight years would be needed to complete the improvements, Atterbury said, at an estimated cost of $100 million. This figure at first appeared to be prohibitive, he admitted. However, when compared with the $70 million which the Pennsylvania had been spending annually on capital improvements in recent years, an additional $10-15 million per year for electrification did not seem unreasonable. Moreover, expenditures for electrification would render unnecessary certain improvements which would otherwise have been made —enlarging the coaling wharf at Thorndale (near the junction of the main line and the western end of the Trenton Cut-Off), for example.[41]

"The immediate factors which influenced the decision to proceed with electrification at this time," Atterbury explained, were as follows:

1. The greater economy of electric traction as compared to steam operation in dense traffic territories;
2. The growth of the southern passenger business;
3. The increasing density of both freight and passenger business on our eastern lines and the probability that in the future more rapid movement would be required;
4. The desirability of utilizing the advantages of electric traction in connection with the construction of our new passenger terminals at Philadelphia, Pa., and Newark, N.J.;
5. The desirability of building a locomotive that would meet the requirements from the standpoint of weight of train, speed, and reliability

that we believe will have to be met in this territory in the next twenty years;

6. The probability that the project, if started now, would be completed with a less total expenditure, all matters considered, than if started at a later date.[42]

The PRR president expected electrification to reduce the number of freight trains run in the corridor by 50%, to eliminate the need for double-heading and the second sections on passenger trains, and to provide cleaner, faster service all around.[43]

When Atterbury proclaimed that "this electrification will exceed in magnitude and in importance that of any other railroad in the world,"[44] he was not exaggerating. The project involved erecting catenary over 325 route-miles and 1,300 track-miles of main line. On a normal workday, the PRR operated an average of 236 passenger trains and nearly 100 freight trains between New York and Philadelphia. Through passenger service between these two cities would consume 238 million kilowatt-hours of electricity annually, the railroad predicted. Philadelphia suburban operations would require an additional 96 million kilowatt-hours per year, and New York–New Brunswick locals would use 11 million kilowatt-hours during the same period. Freight operations would demand still another 137 million kilowatt-hours every year. Once the original electrification was completed, the PRR hoped to extend catenary to the nation's capital just as soon as financially possible. Ultimately the railroad envisioned electrifying its main line all the way to Pittsburgh.[45]

Reaction to the Pennsylvania's electrification plans was almost entirely favorable. The trade weekly *Railway Age* praised Atterbury's announcement as "a declaration of faith in the future development of traffic, of the effectiveness and flexibility of alternating current traction systems, and the ability of the railroad by improved facilities to compete with other forms of transportation."[46] The editors of *Electrical World* (the leading publication of the utility industry) hailed the impending conversion to electric traction as a "great step forward" not only for railroads but also for commercial power companies. The Pennsylvania Railroad and the Philadelphia Electric Company had proven that railroads and utilities could arrive at mutually satisfactory power contracts of major proportions. Only about 1.1% of some 250,000 route-miles of steam railroads in the United States had been electrified thus far. *Electrical World* now urged other utilities to follow Philadelphia Electric's example and begin exploiting this virtually untapped market.[47]

The *Philadelphia Inquirer* congratulated the PRR on its momentous decision and decreed, "This is the Age of Electricity!"[48] The *New York Times* claimed to see political overtones in the timing of Atterbury's announcement. It came just five days before election day. On

November 6, voters would choose either the Republican, Herbert Hoover, or the Democrat, Alfred E. Smith, to be their next President. Many businessmen, betraying their traditional distrust of the Democratic party, believed the contest to be a choice between continued prosperity and an economic recession. The *Times* editorialized that the PRR's decision to disclose its electrification venture before election day showed that there was no basis for labeling Smith an antibusiness candidate.

> Were there the slightest belief in the minds of those responsible for the PRR that any result of next Tuesday's voting could affect the fundamental business structure of America, they would have waited a week before determining to enter into this vast plan . . . The time is not so far past when great corporations like the PRR were willing to help Republicans in their wolf-cries around election time. The action of the PRR leads the great industries of this country to a better plan and abolishes the former evil alliance between big business and Republican politics, for mutual favors rendered.[49]

The *Times* selected Atterbury for special praise. The PRR president had displayed a fine sense of political neutrality, it remarked, even though he was a member of the Republican National Committee and an active supporter of the Republican party within the state of Pennsylvania.[50]

The *Commercial and Financial Chronicle,* a highly respected voice of the nation's financial community, sounded a note of caution amid the accolades. Reflecting the conservatism of many bankers, the *Chronicle* thought that perhaps the Pennsylvania had been too hasty in its move to electrify. Present machinery such as steam locomotives and related facilities had not lived out their economic lives.

> It would seem that since present machinery is adequate, the destruction of perfectly serviceable material will be compulsory and costly. Just because some genius turns out a new machine, must it be bought before it can be paid for, or used before it is needed, or put in the place of another that is yet giving good service?[51]

The *Chronicle* was also worried that with technology advancing at a rapid pace, the PRR's electrification would be obsolete in just a few years. "Revolutionary modifications in this form of motive power may occur," warned the paper. "There is the chance that some new discovery will render useless these improvements before they are finished."[52]

Criticism of electrification which might have been expected from another quarter never materialized, however. The *Locomotive Engineers Journal,* national organ of the Brotherhood of Locomotive Engineers, carried only a brief editorial announcement of the PRR's electric traction plans. After quoting that portion of Atterbury's remarks which dealt with the elimination of double-heading and the running of fewer trains, the *Journal* merely remarked that "a young man working in the territory between New York and Wilmington as a fireman and expecting promotion has a poor outlook."[53] In a later editorial, written as the first stage of electrification neared completion, the *Journal* manifested a fatalistic attitude, even while acknowledging the imminent loss of many jobs. It stated, "The trend of the times seems to indicate quite clearly the passing of the steam locomotive, but it is in the line of march of progress, and a part of the times. It is another product of the machine age."[54] The Brotherhood of Locomotive Firemen and Enginemen, which had numerous members in the employ of the Pennsylvania, never saw fit even to express an opinion in its publication, the *Brotherhood of Locomotive Firemen and Enginemen's Magazine*, regarding the railroad's conversion to electric traction.

The most hostile response to the PRR's electrification came from the National Coal Association. It was disturbed on several counts. The Pennsylvania Railroad was the nation's largest purchaser of coal. During the 1920s, it was buying over 17 million tons annually, or about 3.3% of the national output, mainly for use in its locomotives. A long-distance electrification would substantially reduce the railroad's consumption. The coal industry also feared that a large-scale electrification would encourage on-line industries, which now burned coal, to make similar conversions to electricity, a course of action which would further cut into the demand for coal. Finally, since the Pennsylvania hauled more coal than any other railroad (about a half billion tons every year, or 10% of the national output), it would only be cutting its own throat, economically speaking, if it switched to electric motive power.[55] Well before the PRR committed itself to long-distance electrification, the coal industry tried to convince the railroad not to abandon the steam locomotive. *Coal Age*, for example, exhorted the railroad to experiment with the use of automatic stokers (long anathema to PRR management) and pulverized coal (which only electrical utilities could burn at that time) in the interest of boosting the efficiency of the steam engine. Such improvements were "so much less revolutionary, expensive, and questionable than electrification," remarked *Coal Age*.[56]

At the request of the electric power industry, the Pennsylvania rebutted the charges of the National Coal Association. The railroad agreed that it would require less coal in the future. It did not dispute the coal producers' contention that many on-line industries would soon substitute electricity for coal furnaces. However, the PRR asserted that the coal industry failed to consider that the resultant growth in the demand for electric power would translate into a greater demand for coal, since the vast majority of commercial generating stations were coal fired. The Pennsylvania projected, therefore, that the amount of coal it hauled would eventually increase, not decrease.[57]

The Pennsylvania had determined to electrify its eastern corridor operations only after expending much time and money on preliminary studies. Once the decision was made, there would be no turning back. As if to symbolize its resolve, the PRR allowed the proposed relief line—the Pennsylvania and Newark—to die a quiet death, punctuated only by the demolition of the partially completed bridge at Trenton in 1932.[58] It was well that the railroad had been so deliberate in its investigations, for its commitment to electrification was to be tested severely in the years ahead.

6

···

Frustrations and Triumphs

When President Atterbury divulged his company's intention to electrify, he could not have predicted that less than a year later the stock market would crash, plunging the railroad and the nation into the worst economic depression in modern times. Nor could Atterbury have foreseen the difficulty which the PRR would encounter in obtaining a truly satisfactory locomotive for service on its electrified lines. Numerous other problems presented themselves during the course of the electrification work, but none were so onerous as these two. On the one hand, the Pennsylvania had to fight for its economic life, and on the other, it had to push forward an expensive program of technological improvements.

The railroad planned to complete the electrification by stages, spreading the work over seven or eight years in order to lessen the financial strain of the project. Converting the four-track line between Philadelphia and Wilmington to electric traction, having been contemplated previously as part of the Broad Street terminal improvements, was accomplished in 1928. The line running north from Philadelphia to Trenton was to receive catenary next, along with the 15 miles of track stretching between Sunnyside yard on Long Island and Manhattan Transfer.[1]

The New York zone, then equipped with direct current, presented the most formidable obstacles and therefore the greatest potential for construction delays of any segment to be electrified. The PRR wanted to begin work there at once. If any serious setbacks did occur, the railroad would have plenty of time to deal with them before electrification of the remainder of the line to Philadelphia was completed. Late in 1929, railroad work gangs began preparing to rip out the third rail and install an a.c. system. When the six tunnels (two under the Hudson River and four beneath the East River) had been built, consideration had been given to the use of alternating current. Provision had thus been made for adequate overhead clearance, and supports had been implanted in the crown of the tunnel arches from which a contact wire could be suspended. Unfortunately, because of the increased size of rail used after 1910 and the gradual accumulation of additional stone ballast over the years, the railroad found that the required 15 feet, 8 inches between the rail top and the overhead supports could not be attained. The railroad decided that the only means by which the necessary clearance could be achieved was through removal of some of the ballast.

A crew of about 100 men began this complicated task in 1930. They first jacked up the rails several inches, then removed a portion of the ballast, repositioned the ties, and finally lowered the rails to their new depth. The entire process involved hand labor, owing to the restricted space within the tunnels and the limited ventilation available. The heavy traffic through the bores (an average of 748 trains every day) meant that work could proceed on only one tunnel at a time and only during the late night and early morning hours. Even then, the actual rail raising and lowering operations could take place only between 2:00 and 5:00 A.M., at which time all trains were prohibited from using the tunnel. It was by far the most tedious, time-consuming chore involved in the electrification program. Not until 1932, after the railroad had removed over 20,000 cubic yards of ballast, were the tracks ready for a.c. service.[2]

The third rail had to be retained in the East River tunnels, of course, since the Long Island's trains still used direct current. The PRR had to be extremely careful in seeing that the new a.c. wiring did not interfere with or damage the older d.c. equipment. The railroad also had to take pains to ensure that the fourth harmonic present on the 25-cycle did not combine with the currents of the signal system to produce improper signal action anywhere between Sunnyside and Manhattan Transfer.[3]

The Pennsylvania's own engineering forces performed most of the work in the New York zone. Elsewhere the railroad hired private construction companies. Philadelphia-based Day and Zimmerman, for example, a large engineering and contracting firm having extensive experience in utility construction, did much of the catenary erection. To design the entire electrification and to supervise the work of the contractors, the PRR engaged its longtime consultants, Gibbs and Hill.[4]

The first step in electrifying the main line was to put underground all signal and communication lines in order to avoid electrical and structural interference with nearby transmission and contact wires. The railroad had begun burying its signal lines on the New York division in 1925, partly in anticipation of electrification (although the PRR refused to confirm this) and partly to reduce their vulnerability to the weather. To prepare for the heavier locomotives and higher train speeds which electrification would bring, the PRR replaced the

130-pound rail found in most of the main-line running tracks with 152-pound rail. The heavier rail could sustain 100,000-pound axle loadings at 100 miles per hour, compared with the 80,000-pound at 80 miles per hour limitations of the lighter rail.[5] Higher speeds also necessitated the realignment of the railroad's block signal system west of Manhattan Transfer. Signals had to be respaced to give a two-block indication, with a provision for three-block indication in the event train speeds soared even higher in the future. Position lights similar to those first used on the Paoli electrification were substituted for the remaining semaphore signals, except in the New York terminal area where the original color lights were retained. To further enhance the safety of the new system, the PRR intended to use cab signals in all of its new electric locomotives. The engineman would be able to see in his cab the same position light indication that appeared on the upcoming block signal.[6]

Clearance problems existed in areas other than the New York tunnels, although not with equal severity. The Pennsylvania had to raise numerous bridges between New York and Philadelphia to obtain sufficient space for the passage of the contact and messenger wires. In the Trenton area alone, one railroad and six highway bridges plus a canal aqueduct had to be lifted. All work was done at considerable expense to the PRR. Raising the aqueduct, for instance, meant that the railroad also had to pay for the installation of a new lock and a half mile of new levees to compensate for the elevation in the canal's water level. All bridges, whether raised or not, were screened in at the sides to discourage anyone from showering the tracks below with debris—especially with debris that conducted electricity.[7]

While the Pennsylvania left most of the actual construction work to the contractors, the railroad's own engineering corps developed much of the equipment which the contractors were to use. To speed up the laborious task of pouring the foundations for catenary poles and guy anchors, the PRR provided the contractors with specially built five-car concrete trains. These trains (five were in use at the height of construction) allowed cement, water, sand, and gravel to be mixed on board and then poured directly from the mixing car. Each train set was fitted with floodlights and steam heating pipes for 24-hour, year-round operation. To maximize the efficiency with which the catenary poles were emplaced and to reduce interference with passing trains, the railroad's engineers invented a unique type of car-mounted, steam-driven crane. Only the hoisting gear and the operator's cab rotated; the propulsion machinery remained stationary. This arrangement eliminated the obstruction caused by the rear end of the rotating crane overhanging an adjacent track. The third and final stage of catenary construction—after the foundations had been poured and the poles erected — involved the stringing of the messenger, contact, and other supporting wires. Here again the Pennsylvania's technical ingenuity aided the contractors while keeping interference with the normal flow of traffic to a minimum. The railroad devised a special kind of boxcar having "outriggers" or platforms extending out from the top of the car. Workers could thus hang wires above three tracks from a single car without delaying trains on adjoining tracks. Whatever electrification's eventual benefits, the Pennsylvania was not about to permit its installation to impair the railroad's service to passengers and shippers in the interim. And because conversion to electric traction was such an expensive proposition, the PRR maintained a close surveillance of the contractors' work through all phases of construction. It aimed to make sure that the job was done efficiently and in a manner which would keep future maintenance and repair costs to a minimum.[8]

The Pennsylvania likewise worked closely with the locomotive manufacturers. In the same statement of November 1, 1928, which heralded the forthcoming electrification, President Atterbury noted that "with the assistance of the electrical companies we have developed three types of electric locomotives which . . . will provide any power demanded by size, weight of trains, or speed."[9] At that time, Westinghouse and General Electric were progressing independently toward the development of a traction motor coupling high horsepower with compact, lightweight design. In 1927, Westinghouse had offered the PRR a basic design which the manufacturer claimed would fulfill the railroad's requirements, but it needed further refinement before becoming a full-fledged production model. Thus Atterbury's statement regarding the status of his road's electric engines was a bit premature. It did demonstrate, however, that the Pennsylvania's motive power planners had a clear conception of the kinds of locomotives they wanted well before the time arrived to put these machines into actual operation. By 1930, both Westinghouse and General Electric had developed what each believed to be a suitable traction motor. Westinghouse produced the Type 425 motor, and GE offered the nearly identical Type 617 (which soon became the basis for an improved Type 625).[10]

In charge of the engineering phase of the PRR's electrification program in 1928 was a 46-year-old former Vermonter named John Van Buren Duer. After receiving a mechanical engineering degree from Stevens Institute of Technology in 1903, Duer had worked briefly as an electrical equipment inspector for both General Electric and Westinghouse before taking a similar job on the Long Island Rail Road. When the PRR in 1910 was looking for experienced electrical personnel in preparation for the inauguration of its New York terminal service, Duer was hired as a foreman of motormen. He

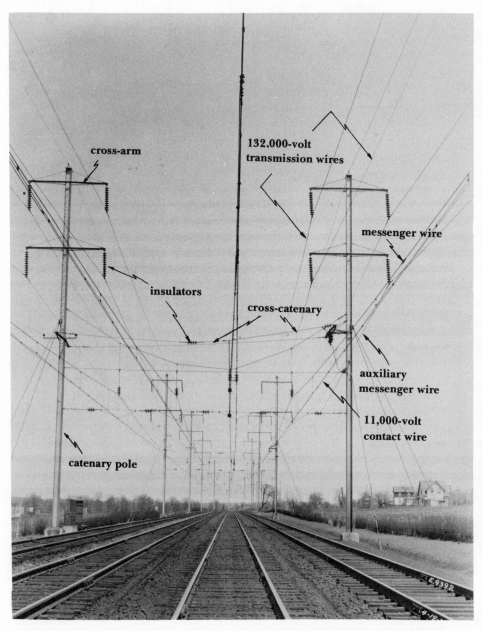

Main components of a typical section of catenary. (Altoona Public Library)

worked in New York for only three years before being promoted to assistant electrical engineer at Altoona. In this new post, he worked closely with such eminent Pennsylvania mechanical engineers as Alfred W. Gibbs, Axel S. Vogt, and William F. Kiesel, Jr. In 1920, the railroad named Duer "Chief Electrical Engineer," a newly created position which indicated how seriously the PRR regarded electrification. In this capacity, Duer helped to design the ill-starred L5 and directed most of the studies of electrification which the PRR had done during the 1920s.[11]

Duer and his staff patterned their blueprints for electric motive power after the designs of the PRR's steam locomotives. The railroad was striving to obtain the maximum horsepower per axle consistent with tractive effort requirements. It wanted locomotives which it could duplicate in large quantities so that production and maintenance costs could be held to a minimum. To design and build these locomotives would be no mean feat. Not only did the new traction motors have yet to prove their worth; electrification itself had made such meager inroads among American steam railways that mass production of anything was practically unheard of.

That the PRR therefore should follow certain criteria set by its steamers was understandable. In fact, the railroad developed its three electric types specifically to replace on a class-for-class basis the three types of steam engines most prevalent on the main line in the territory to be electrified. In the case of the most powerful new electric class, the P5, Duer explained:

> Consideration was given to the horsepower and tractive effort of a steam locomotive that was handling through passenger trains successfully; it was felt that if an electric locomotive could be built with the same or better horsepower and tractive effort than that of this particular steam locomotive, the resulting electric locomotive would provide a satisfactory unit for handling through passenger trains.[12]

The steam locomotive to which he referred was the 4–6–2 K4s, which at that time was pulling limiteds of up to 18 cars over the New York division. The PRR intended that the P5 handle the same train weights but at the higher speeds which were necessary to take maximum advantage of electrification. In a similar manner, electric Class O1 was to assume the duty of hauling shorter and lighter passenger trains, an assignment then held by E-class 4–4–2 steam locomotives. The L6 electric was to replace the 2–8–2 L1s steamer as a general utility freight locomotive.[13]

The O1 represented essentially a shorter and less powerful version of the P5. Both classes employed the same type of 1,250-horsepower twin motors driving each axle through a geared quill mechanism. In this arrangement, the motors, positioned between the wheels but at a

Builder photographs of L6 No. 7825, O1c No. 7857, and P5 No. 7898. (Altoona Public Library)

considerable distance above the axle, were geared to the hollow quill surrounding the axle. A form pioneered by Westinghouse and the New Haven, it combined the advantages of both the gearless quill and the geared drives. The P5, with its 2–C–2 wheel arrangement, had an output of 3,750 horsepower, while the 2–B–2 O1 was rated at 2,500 horsepower. Each class rode on 72-inch drivers. (The K4s and the various E subclasses had 80-inch drivers, but after the failure of the L5's, which also used 80-inch driving wheels, the PRR decided to apply a slightly smaller size wheel to its new electrics.) Both the P5 and O1 sported the same black, utilitarian box cab configurations. The L6 bore an external resemblance to the other two classes, but internally it harbored several differences. In place of the twin motors and geared quill drive, the L6 used only one motor per axle and a direct-geared (that is, no quills), axle-hung drive mechanism. Having a 1–D–1 wheel arrangement, it rode on eight 62-inch driving wheels, identical in size and number to those of the L1s.

The Pennsylvania insisted that as many parts as possible be interchangeable, regardless of which manufacturer might build the locomotives or their components. This was a stipulation that had a precedent in the L5; however, the PRR had now progressed to the point where it desired maximum interchangeability among all three classes, not merely within a single class. In another attempt to reduce long-term costs, each locomotive type had large roof hatches through which internal machinery could be lifted in entire units for repair. Similarly, whole driving wheel assemblies could be dropped from below.[14]

In 1930, the Pennsylvania outshopped its first pair of O1's, both of which were equipped with Westinghouse electrical gear. The railroad immediately put them into experimental service on the Philadelphia-Wilmington and the newly electrified Philadelphia-Trenton runs. By the end of 1931, the first two had been joined by six more O1's. General Electric had supplied the electrical components for three of these units, American Brown-Boveri two, and Westinghouse one. To distinguish them from the original units, these six were variously classed O1a, O1b, and O1c. Their primary distinction lay in the size of their traction motors, so that the subclasses' horsepower rating ranged between 2,000 and 2,500. As the eight locomotives began their trials, the PRR turned its attention to the P5. It began construction of two at Altoona in the spring of 1931.

Confident that the new machines would meet expectations, the railroad announced in June 1931 that it had ordered a total of 90 of these engines. The 2 then being built at Altoona were included in the order and would serve as test-beds for the remaining 88 units. The Pennsylvania itself would construct 11 P5's using GE and Westinghouse electrical gear. General Electric contracted to build 25 at its

Erie, Pennsylvania, plant. Westinghouse and its partner in the railway electrification business, Baldwin, received an order for 54 locomotives. Besides interchangeable mechanical and electrical parts, all the P5's were to have 90-mile-per-hour gearing for high-speed passenger work. To accommodate its large order, Westinghouse had to institute a traction motor assembly line at its East Pittsburgh works for the first time in the 30-year history of its heavy electric traction program. From the Westinghouse plant, the electrical components were shipped to Eddystone (near Philadelphia), where Baldwin —which built the car bodies and mechanical parts—did all the final assembly work.[15]

In August 1931, the first two P5's joined the O1's for in-service testing. The PRR engineers quickly noted certain minor defects—the traction motor blowers were too small, for example—and took steps to correct these difficulties in the 88 locomotives which were to follow. These units, the delivery of which began in mid-1932, were subsequently designated by the railroad as P5a's.[16]

The Pennsylvania followed a similar procedure with regard to the L6. J. V. B. Duer proclaimed this locomotive to be "the best all around unit for handling our freight trains." In the fall of 1931, the railroad built 2 L6's at its Altoona shops. Even before they had left the erecting bay, however, the PRR had sufficient faith in the L6's to place an order for 30 more with the Lima Locomotive Works of Lima, Ohio. The total cost of the L6 units (exclusive of electrical gear, which GE, Westinghouse, or American Brown-Boveri would provide) was $1,350,000—Lima's largest order for 1931.[17] The selection of Lima was a curious one. While it was one of the "big three" steam locomotive builders (with Baldwin and Alco), it had practically no experience with the construction of electric locomotives. Moreover, the PRR had never dealt much with Lima even where steam engines were concerned. The railroad had traditionally awarded the bulk of its outside orders to Baldwin. Possibly the L6 order was an attempt by the PRR to spread more evenly among the builders what little money existed for new equipment during these depression years. Lima was also an important on-line customer.

The P5, O1, and L6 classes were not the only new motive power acquired by the Pennsylvania in connection with expanded electrified operations. In 1926, the railroad's Altoona shops built a class of electric switching locomotives for service in the New York and Philadelphia terminal areas. At first, these 0–6–0 units were semipermanently coupled in pairs and designated Class BB1 (a.c. equipped), BB2 (d.c. equipped), or BB3 (a.c. equipped for operation by the Long Island Rail Road). After several years, all the pairs were detached and the BB2's converted to a.c. propulsion to make them compatible with the a.c. equipment then being introduced into

A B1 under construction at Altoona, 1934. (Altoona Public Library)

Pennsylvania Station. All the units were also given a new class, B1. Fourteen more B1's were delivered from Altoona in 1934–35, bringing the PRR's total to 28 units, not including the 14 B1's in service on the Long Island. The B1's made the Pennsylvania the owner of by far the largest number of electric switching locomotives of any American railroad.[18]

Unfortunately for the railroad, the headway which it was making in the realm of engineering was not matched by similar progress on the financial front. The Pennsylvania had hardly awarded its first contracts when Wall Street suffered the calamitous tumble of October 1929. Although the crash instantly sent waves of uneasiness rippling through the business community, the railroad gave no indication that the stock market's misfortune would affect electrification plans. In a speech to the Manufacturers Club of Philadelphia on December 7, 1929, President Atterbury brushed aside fears that a recession might curtail his company's capital improvements plans. He stated, "We have every intention of going ahead with our entire improvement program as originally planned, without any slackening, retrenchment, or postponement. . . . We believe with President Hoover that the country and its fundamental business conditions are sound."[19]

The year 1930 saw the economic downturn of the last quarter of 1929 become more intense. The PRR's passenger and freight revenues declined markedly. Net income for 1930 sank to $68 million, down $33 million from the record high of $101 million posted for 1929. On the other hand, the picture was by no means a bleak one. The Pennsylvania cut railway operating expenses by 13% in 1929. It still managed to record a respectable 72.5% operating ratio for the year, up only 2% from 1929. Its net income for 1930, while off sharply from the previous year, exceeded net earnings for every year prior to 1928. And of most importance for the electrification project, the Pennsylvania had little trouble selling $60 million worth of 44-year, 4½% bonds—its first bond issue since 1921.[20]

The recession soon evolved into a full-blown depression. Some PRR shareholders expressed anxiety about spending huge sums of money during a time of plummeting revenues. Atterbury tried to allay these worries in a letter which accompanied the stockholders' autumn dividend for 1930.

> The prudent use of new capital primarily leads to further operating efficiencies and economies. Capital expenditures must also be made for the more obvious purpose of constantly improving the property and providing satisfactory service, so that traffic may not only be retained but increased when business shall again resume its normal progress, which it must do. The PRR will then be in a better condition than ever to serve the needs of a prosperous nation.[21]

In fact, Atterbury believed that the current rate of improvement expenditures was insufficient. He proposed that spending be increased and that territory to receive electrification be enlarged. The board of directors concurred. On February 17, 1931, President Atterbury announced that the PRR would extend catenary beyond Wilmington to Baltimore and Washington. Whereas the railroad had not expected to complete the original electrification work until 1935, it now intended to step up spending so that the entire project (including the extension to Washington) would be finished late in 1934. Atterbury said that his company planned to spend nearly $175 million on electrification over the next two and a half years.[22] In a detailed statement to the press, he carefully explained why the PRR's management had chosen such a bold course of action.

> It is our view that commodity prices are now at a level, and the efficiency of labor is so great, that these improvements can definitely be contracted for on an exceptionally favorable basis. Furthermore, at a time like the present, with reduced traffic, the work can be done with much less interference from passing trains, which, of itself, would constitute a definite economy. . . . The making of these improvements will obviously involve new financing, but the Pennsylvania Railroad, in view of the existing low interest rates and the plethora of savings in the country that are clearly available for investment, as soon as confidence is restored, regards the

present and the near future as a most favorable occasion for the necessary financing.[23]

Such an accelerated program of electrification was evidence that the PRR was not paying mere lip service to President Hoover's declarations about the eventual return of prosperity. "The Pennsylvania Railroad considers that now is the time to express its confidence in acts," affirmed Atterbury.[24]

Atterbury foresaw no insurmountable difficulties in raising the necessary money to complete the electrification. He said that the railroad would be able to finance the remaining work through its own earnings, plus bond issues and long-term bank loans.[25] On March 12, 1931, the PRR disclosed that it had just sold another $50 million worth of 4½% bonds. The cash which the transaction generated, coupled with the issuance of equipment trust certificates to cover the cost of new locomotives, would be enough to sustain electrification work, the railroad estimated, for another 18 months.[26]

With catenary between Philadelphia and Trenton and between New York City and New Brunswick energized in 1930, the Pennsylvania proceeded to close the 25-mile gap between Trenton and New Brunswick. Crews also began preparing the main line from Wilmington to Washington for electrification, although catenary erection was not scheduled to begin until the following year. The railroad and its contractors had over 6,000 men laboring on these two segments. In April, the PRR purchased 47,500 tons of fabricated steel for poles, beams, braces, crossarms, and other components of the catenary system, a sufficient quantity to keep the men at work for the rest of the year.[27]

As 1931 drew to a close, however, the PRR realized that in order to purchase the materials necessary to finish electrification, it would have to issue more bonds. Atterbury and his lieutenants had tried valiantly to avoid returning to the bond market so soon. They had redoubled their economy drive in 1931 in hopes of financing electrification from their company's earnings. Effective July 1, 1931, for example, officers' salaries had been slashed by 10%. Unionized employees agreed to take a similar reduction in pay beginning February 1, 1932. The downward trend in traffic permitted the railroad to lay off many thousands of its employees, thereby saving additional funds. Deferred maintenance continued. As a result of these and other stringent measures, railway operating expenditures for 1931 showed a decline of $74 million or 17.4% from 1930 totals. Unfortunately, net income had fallen at a more rapid pace than the PRR's ability to shed unnecessary expenses. At $19.5 million, net income for 1931 was a mere fraction of predepression levels. The railroad had little choice, therefore, but to prepare another bond issue. In contrast to the successes it had enjoyed in February 1931, the PRR in late 1931 and early 1932 could find no buyers for its bonds. This financial reverse came in spite of the fact that the Pennsylvania Railroad was one of the few American rail lines still operating in the black and still paying dividends.[28]

The railroad did not abandon hope of getting the needed cash. In March 1932, it applied for a $55 million loan from the Reconstruction Finance Corporation (RFC). Congress had created the RFC only two months earlier to lend money primarily to banks, insurance companies, and (with the consent of the Interstate Commerce Commission) railroads. Loans were to be made only to those firms which were still solvent and could provide adequate collateral but which could not, for one reason or another, obtain money from the usual private sources. In making application for the money, Vice-President A. J. County warned that if the loan were not approved, all electrification work would quickly come to a halt. He estimated that the Pennsylvania itself could spend but another $13 million on electrification in 1932, about one-tenth of the sum it had originally budgeted for that work.[29] The RFC's board of directors, while desiring that the PRR continue the conversion to electric traction, balked at lending it such a large amount of money when the railroad would be spending only a relatively small quantity of its own funds. President Atterbury then suggested that if the RFC lent the railroad half its original request—$27.5 million—the PRR would somehow raise a like amount from private resources. The RFC's directors readily agreed. On May 19, 1932, after the Interstate Commerce Commission had given its formal consent, the Corporation approved the loan for three years at 6% interest.[30] On the strength of this sanction of electrification by the federal government, the railroad was able to gain matching funds primarily in the form of long-term bank loans. Prior to RFC's action, banks had been unwilling to make loans to the PRR for periods longer than 90 days.[31]

The injection of RFC money enabled the Pennsylvania to continue to advance electrification throughout 1932. Construction of the Philadelphia terminal improvements was nearing completion by this time. Only the massive Thirtieth Street Station remained unfinished. The railroad had moved its executive offices from the rapidly deteriorating Broad Street Station to the upper floors of the new Suburban Station in July 1930. (Lower floors were rented as office space to commercial tenants.) On September 28 of that year, m.u. trains began running to and from the new station. Protracted negotiations with the Philadelphia Rapid Transit Company concerning removal of their elevated tracks along Market Street had delayed construction of the larger and more expensive structure at Thirtieth Street until after the onset of the depression. A shortage of funds had then

Artist's rendering of P5a No. 4763 on the upper level platform at Philadelphia's Thirtieth Street Station. (Pennsylvania Historical and Museum Commission)

further hindered progress on the new station. By late 1932, however, the upper or suburban portion of Thirtieth Street Station was ready for use. The PRR expected to open the rest of the station, including the spacious main concourse and the platforms for through trains, sometime in 1933.[32]

Even before the approval of financing by the Reconstruction Finance Corporation, the railroad had concluded arrangements for the supply of electric power. The PRR's own Long Island City generating station would furnish electricity for operations in the New York zone. The Philadelphia Electric Company would provide power from Newark at least as far south as Wilmington. The utility fed current to the railroad at two locations adjacent to two of its coal-fired plants: Richmond, in northeast Philadelphia, and Lamokin, in Chester. Philadelphia Electric's Arsenal Bridge substation, originally built for the Paoli electrification, was retained as an auxiliary connection. Negotiations were underway with Philadelphia Electric and other utilities with regard to supplying power south of Wilmington and west along the Low Grade Line.[33]

In designing the electrical distribution and transmission system, J. V. B. Duer reported that "economy of first cost and operating

Eastern facade of Suburban Station, Philadelphia. (Conrail)

maintenance, as well as reliability of operation and simplicity of detail, were leading points receiving consideration." Representatives from the PRR, Philadelphia Electric, and the electrical manufacturers (especially Westinghouse, which received the bulk of the railroad's orders for trackside electrical gear) worked in a cooperative effort to achieve these goals. The Long Island City station posed few problems. It generated single-phase, 25-cycle alternating current at 11,000 volts—identical in every respect to the current used in the contact wires. The single-phase generating capacity of the Philadelphia Electric Company, on the other hand, was limited. The use of huge rotary converters was therefore necessary to convert the 60-cycle, three-phase power of the utility to 25 cycles, single-phase. Philadelphia Electric had already installed this equipment at the Lamokin substation in conjunction with the electrification to Wilmington in 1928. At Richmond, the utility installed additional converters similar in design to those at Lamokin but having increased

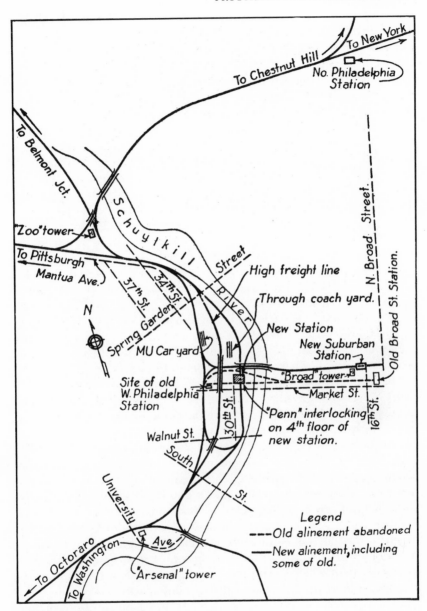

To Chestnut Hill

To New York

No. Philadelphia Station

To Belmont Jct.

Schuylkill River

Street

"Zoo" tower

To Pittsburgh
Mantua Ave.

37th St.

34th St.

N. Broad Street.

Old Broad St. Station.

N

Spring Garden

High freight line

Through coach yard.

New Station

New Suburban
Station

MU Car yard

Site of old
W. Philadelphia
Station

30th St.

"Broad" tower

Market St.

"Penn" interlocking
on 4th floor of
new station.

16th St.

Walnut St.

South

St.

University

Ave.

Legend
- - - Old alinement, abandoned
——— New alinement, including
        some of old.

To Octoraro

To Washington

"Arsenal" tower

PRR Philadelphia terminal area. (*Railway Age*)

output. They were in fact the largest rotary converters built to that date.[34]

In accordance with the Pennsylvania's quest for maximum reliability at minimum cost, electrification as implemented in the early 1930s incorporated numerous improvements over the original installation of alternating current on the Paoli and Chestnut Hill lines. The railroad was now using a much higher voltage for transmission purposes. Transformers stepped up the 13,200-volt current coming from Philadelphia Electric's substations to 132,000 volts before feeding it into the principal transmission lines. Such a high voltage allowed the railroad to use materials lighter in weight and greater in conductive efficiency than those found in the 44,000-volt Paoli electrification.[35] The higher voltage also permitted PRR power dispatchers to more easily and quickly move almost any desired quantity of electricity to any point in the 350-mile system. The Pennsylvania saved even more money by having the same catenary structure which supported the contact wires carry the transmission lines, too, thus precluding the need to purchase a separate right-of-way and erect a separate transmission system. Railroad-owned substations positioned at frequent intervals along the tracks then reduced the 132,000-volt current to 11,000 volts and fed it into the contact wires.

The electrification encompassed some of the most heavily traveled trackage in the United States. To guard against a power failure which could paralyze train movement, the railroad contracted with the Public Service Electric and Gas Corporation of New Jersey to supply power on an emergency basis. The utility constructed a substation at Metuchen, near New Brunswick. Through this connection, electricity could be channeled north or south as needed.[36]

The Pennsylvania Railroad's electrification of the 1930s included many other technical improvements not found in earlier a.c. installations. Many of these advances were developed by the railroad or its suppliers specifically for this great project. Westinghouse, for example, introduced in 1925 a high-speed (1/25 second) circuit breaker. The device was tested on various segments of the PRR's Philadelphia suburban routes. Before applying the circuit breaker to the long-distance electrification, however, Westinghouse engineers had to develop a high-speed trolley relay to work in combination with the breaker and to be so highly selective that it could distinguish between an actual fault and merely a load of higher current value. Another special feature was the inclusion of a series of switches at each substation which permitted current to be taken from either direction. In most places, each of the running tracks was paralleled by its own 132,000-volt transmission line. In the event of a power failure in one of the lines, the switches allowed electricity to be taken from another line, even if the current were flowing in the opposite direction. It was an arrangement which made for extremely reliable service. A multitude of other unique engineering problems, such as protecting the electrical apparatus from the potentially devastating effects of lightning, accompanied the implementation of large-scale electrification. Each demanded and received the careful attention of

A typical substation along the New York-Washington mainline. (Altoona Public Library)

the engineers, who often countered with solutions as unique and specialized as the problems themselves.[37]

On December 9, 1932, the PRR inaugurated multiple-unit car service between Penn Station in Manhattan and New Brunswick. The only portion of the New York division which remained to be converted to electric traction was the segment between New Brunswick and Trenton. Several weeks later, the railroad announced that this stretch, too, had now been fully energized. The Pennsylvania expected to begin operating electrically powered passenger trains between New York and Wilmington in the near future, pending extension of the catenary to Baltimore and Washington.[38]

Early on the morning of Monday, January 16, 1933, a small crowd gathered in the depths of Pennsylvania Station to commemorate the departure of the first electric train to Philadelphia. New York's

Mayor John P. O'Brien was on hand along with a delegation of lesser civic notables. George LeBoutillier, vice-president in charge of the New York zone, led a covey of railroad officials. Also present was a shiny new P5a, which the railroad had assigned to pull the special train. Suspended between the locomotive's handrail and a nearby post was a white satin ribbon. At precisely 9:00 A.M., a bugler from the Sixteenth United States Infantry Regiment blew the command, "Forward." Mayor O'Brien snipped the ribbon, and the P5a and its cars glided slowly away from the station platform. On another platform not far away, a band of commuters inbound from Long Island had been watching the ceremonies. "While uncertain of its significance," a *New York Times* reporter observed, "they gave a cheer as the Philadelphia-bound train pulled out."[39]

The whole affair was remarkably underplayed for an event of such significance. It certainly did not capture the public's fancy in the manner that the opening of Pennsylvania Station had 23 years earlier. No large crowds had gathered in celebration. The Long Island commuters, who happened to be present only by chance, had comprised the greater portion of the general public's representation, and they were not fully cognizant of the importance of the occasion. National press coverage was virtually nonexistent except in the trade publications. Neither of the two major New York newspapers awarded the event front page coverage. The PRR did not even have any of its top executives on hand. LeBoutillier was no substitute for Atterbury or any of his high-ranking assistants who were so instrumental in promoting the cause of electrification.[40]

In Philadelphia, ceremonies were conducted in the same anticlimactic fashion. The special train from New York arrived at Broad Street Station at 10:57 A.M. The first electrically powered train to leave Philadelphia for New York departed at 1:03 P.M. amid little fanfare. Philadelphia's Mayor J. Hampton Moore rode in the cab of P5a No. 4764 from Broad Street as far as the first stop, North Philadelphia.[41]

The Pennsylvania initiated long-distance electric service on a step-by-step basis. In this manner, employees had ample opportunity to familiarize themselves with the maintenance and operation of the new equipment. Engine crews for the new electric locomotives were drawn from the regular service pools in the New York-Philadelphia territory. Several days of classroom instruction were followed by a week or so of actual operating experience under the tutelage of Westinghouse and General Electric factory representatives and experienced crews borrowed from the New York zone. The ease with which enginemen and firemen learned the ways of their new electric machines reflected the simplicity of design that characterized electric motive power. At first, many older enginemen found it difficult to overcome their distrust of the new technology and refused the opportunity their seniority afforded them to qualify for electrified service. As a result, the choice assignments on the best trains (which, of course, were the first to be electrified) went to junior men, much to the young men's surprise. However, as word of the "luxury" afforded by the electric locomotive spread through the ranks, the old hands realized the folly of their ignorance and soon displaced the younger enginemen from their comfortable jobs.[42]

No such prejudice against electrification existed among firemen, who naturally had less appealing duties aboard steam engines and were eager to make the transition to the electrics. The outlook of most PRR firemen had changed little since 1911, when one of their number had explained: "It will be a great change for [the fireman] to sit in a nice clean cab, equal to a Pullman coach, with little more to do than keep his eyes open, ring the bell for crossings, and look wise."[43] Surely this enthusiasm would have diminished markedly had the Pennsylvania taken this statement at face value and tried to eliminate firemen from electric locomotives. Happily for the firemen, the railroad made no such attempt. Not until the 1950s did the redundancy of firemen aboard electric locomotives become an issue. Even then, it was a mere corollary to the larger question of the need for firemen aboard diesels.

So that employees could become fully accustomed to their new jobs and equipment could be given a final inspection for safety and reliability, the railroad maintained all newly electrified trains on their former steam schedules. Only New York-Philadelphia passenger trains—"clockers," in the parlance of the traveling public—were hauled by electric locomotives during the first few weeks. Early in February, New York-Washington trains began to receive electric motive power for the run north of Wilmington. Six weeks after that, the PRR started substituting electric for steam power on its fleet of east-west trains. On April 8, 1933, the Pennsylvania's magnificent flagship, the Broadway Limited, made its final steam-powered run west from Manhattan Transfer. Beginning two days later, all east-west passenger trains to and from New York ran behind electric locomotives between Pennsylvania Station and the end of catenary at Paoli.[44]

The electrification of the long-distance passenger trains spelled the end for Manhattan Transfer, a station which ironically owed its birth to electric traction. Surrounded by bottomless marshland, Manhattan Transfer was accessible only by rail and therefore was not a station in the traditional sense of the word. It served simply as the point where the PRR exchanged steam and electric engines, and where passengers made connections between trains of the Pennsylvania and the Hudson and Manhattan. Possessing no architectural significance of note, the station consisted of two long, single-story buildings set on brick platforms and paralleled on either side by a multitude of tracks.[45] Yet it etched itself indelibly into the minds of millions of travelers, who came to remember it as the site from which—whether through a Pullman window or from one of the undulating red brick platforms—they first glimpsed the outline of Manhattan's looming towers. For countless other, more technically minded passengers raised in an age of steam, Manhattan Transfer provided a never-to-be-forgotten introduction to the mystique of electric traction. Few persons who passed through the place in daylight hours failed to remark on the curious setting of the station. A more isolated locale, within sight of and yet seemingly so remote from one of the world's great metropolitan areas, could hardly be imagined.

So renowned had Manhattan Transfer become by the end of its first decade that writer John Dos Passos took the station's name for the title of a novel, published in 1925 and instantly a best seller. While Dos Passos's *Manhattan Transfer* dealt with social rather than railroad themes and most of the action took place in New York City, the book's name was still an appropriate one, as any modern-day reader will discover.

In 1929, as part of the terminal improvements surrounding the Pennsylvania's long-distance electrification plans, the PRR, the H&M, and the city of Newark signed an agreement whereby all three parties pledged to contribute to the erection of a spacious new station along Market Street in Newark. Connections between the Hudson and Manhattan's tube trains and the PRR's long-haul trains were to be made there, rather than at Manhattan Transfer. The depression delayed construction work for a number of years, but in 1937, the Newark Station was finally opened. The Pennsylvania then petitioned the New Jersey Board of Public Utility Commissioners for permission to abandon the Transfer. That agency approved the request in June 1937, and one of the most famous railroad locations in America soon faded into oblivion.[46]

The Pennsylvania initially had hoped to eliminate not only Manhattan Transfer but Penn Station itself as points where motive power had to be exchanged. Both the New Haven and the PRR used essentially the same type of a.c. system. By establishing a pattern of locomotive pooling (for example, by having PRR locomotives run north to New Haven and those of the New Haven travel south to Philadelphia), engine runs would be lengthened. More efficient equipment utilization would be achieved, as locomotives spent more time on the road and less time in the terminals. The New Haven was intrigued by the idea. During the summer of 1929, it loaned one of its EP-2 class locomotives (of 2,640 horsepower) to the Pennsylvania to undergo testing in the Hudson tunnels. While the EP-2 performed satisfactorily even when pulling exceptionally heavy trains, both PRR and New Haven engineers concluded that had the locomotive ever been halted on the tunnels' steep 1.93% westbound grade, it never would have been able to start its train again.[47] A subsequent examination of this problem by the two railroads confirmed these suspicions. Double-heading would overcome the difficulty in the tunnels, to be sure; but it would be wasteful elsewhere. Always mindful of its responsibility to provide its customers with the most reliable transportation possible, the Pennsylvania refused to use but a single EP-2, fearing the disruption in traffic caused by the possible stalling of the locomotive on the sharp grade. The PRR reluctantly discarded the idea of running motive power straight through its Manhattan terminal.[48] Had such an arrangement been effected, it would have preceded similar power-pooling agreements on other American railroads by three decades.

The Pennsylvania's most powerful electric locomotive, the P5, had not progressed beyond the blueprint stage at the time of the EP-2 tests. However, later events would show that even the improved version of this machine—the P5a—would have been little more suited than the New Haven's locomotive to the kind of pooling which the two railroads had under consideration. These same events would also partially vindicate the PRR's decision to greet the commencement of long-distance electrification in a low-key manner. As the year 1933 wore on, the railroad began to experience some embarrassing problems with the P5a—problems so serious that they ultimately forced the Pennsylvania to return to steam power for a time.

Even before the P5a's flaws had become apparent, the PRR had abandoned plans to duplicate the very similar O1, which the railroad found to be of only limited usefulness. The locomotive's most glaring defect was insufficient weight on drivers. Fully half of the O1's total weight of 150 tons rested on its eight nonpowered wheels. As a result the locomotive exhibited a troublesome slipperiness when trying to start any but the lightest of trains. It did perform capably enough on many trains when used in pairs, but this was a practice which defeated one of the main purposes of electrification.[49]

The P5a's shortcomings did not manifest themselves until full electrified service had gotten under way and were more complex than the deficiencies that plagued the O1. With 110 tons on drivers, 55,000 pounds starting tractive effort, and 6,400 horsepower (one-hour rating), the P5a encountered less difficulty than the O1 in starting trains. However, it did have trouble maintaining high speeds on the short but formidable grades in electrified territory—the western end of the Hudson tunnels, for example. The P5a in most instances could still outpace a laboring K4s on these slopes, even with a 17- or

18-car train. In that sense, the electric locomotive brought about a definite improvement. Nonetheless, the Pennsylvania would not be able to attain with the P5a the very fastest intercity schedules unless it reduced the number of cars per train or double-headed the engines. Neither alternative appealed to the efficiency-conscious PRR management.[50]

During the spring of 1933, more serious problems arose to compound the railroad's dilemma. First the P5a's began to display a pronounced lateral motion when traveling at speeds greater than 70 miles per hour, a phenomenon reminiscent of that experienced nearly 30 years earlier when the PRR was developing an electric locomotive for New York tunnel operations. Next railroad mechanical inspectors detected small cracks in the driving axles of some of the P5a's.[51] Pennsylvania officials became alarmed. These new faults not only impeded electric traction's economic fulfillment; they posed grave safety risks as well. The PRR immediately set to work to determine the causes of the high-speed oscillation and the cracked axles.

To deal with the oscillation, Duer and his staff of electrical and mechanical engineers decided to subject the P5a to the same kind of Brinnell hardness test which had been used successfully at Franklinville in 1907. The O1 also was to undergo these tests, since that class produced the same type of sideway as the P5a. Inasmuch as the O1's had already been permanently exiled to light-duty assignments, however, the railroad was not overly concerned about correcting their flaws. A segment of main line near Claymont, Delaware, was selected as the test site and 252 cast-steel ties—the same ties that the PRR had used at Franklinville—were embedded beneath one of the northbound tracks. Every other tie was fitted with a hardened steel ball, whose purpose was to record the lateral thrust of a passing locomotive by impressing itself into a slab of boiler plate. At the Pennsylvania's request, General Electric engineers applied electronic strain gauges to the axle ends of the locomotives being tested. These instruments also measured lateral motion. Westinghouse engineers were on hand with still more electronic measuring devices. The tests began in April 1933 and continued sporadically throughout the rest of the year. The railroad gathered and compared data from a number of P5a's and O1's, plus representatives of other classes of Pennsylvania electric and steam locomotives.[52]

Meanwhile, PRR engineers had diagnosed the P5a's axle cracks as stemming from the severe wrenching or twisting action of as much as 2,150 horsepower per axle. The P5a had an unusually high axle-to-horsepower ratio, even for a locomotive whose owners traditionally sought to incorporate as much horsepower on as few driving wheels as possible. No PRR steam or electric locomotive ever matched this ratio, before or since the P5a. The excessive torque which this combination produced placed an extraordinary amount of strain on the relatively narrow axles. Probably contributing to the axle stress was the utilization (by both the P5a and the O1) of a single-ended, geared quill drive. In this drive mechanism, only one driving wheel on each axle actually received power (hence the term "single-ended"). The front and rear axles were powered on the right side, the middle axle on the left. The unevenness of the power application which accompanied the use of single-ended drives, when coupled with the burden of more than 2,000 horsepower, could only have further intensified the strain on the axles.[53]

By the fall of 1933, the Pennsylvania's engineers believed that they had devised satisfactory solutions to the P5a's mechanical defects. On October 28, the railroad announced that it was temporarily withdrawing its fleet of sixty or so P5a's from service. Company spokesmen refused to comment on the rumors that had been circulating regarding the locomotives' thus far disappointing record. The PRR would say only that the P5a's had been sent to the road's Wilmington shops "in order to effect minor mechanical improvements which, in the light of experience, will result in safer and more efficient operation."[54] The reliable K4s steamers, so unceremoniously expelled from their main-line duties just a few months before, would replace the electrics. The decision to return to steam power, even if just for a few weeks, must have been made only after a good deal of anguished deliberation in the road's new general offices. But whatever embarrassment the railroad might suffer, General Atterbury was not admitting defeat. Rather he was just conducting a necessary, if somewhat humbling, strategic retreat.

The Claymont tests had revealed that deficiencies in the P5a's suspension system bore the major blame for the locomotive's poor tracking qualities. The two engine trucks did not incur sufficient resistance at high speeds to prevent them from shifting from side to side. To alleviate this difficulty, the PRR first redistributed the weight of the locomotive, reducing driving axle loads by 10,000 pounds per axle and correspondingly increasing truck axle loads by 7,500 pounds each. The heavier load on the trucks reduced their oscillative tendencies. To further limit oscillation, the railroad fitted the center pin of each truck with lateral resistance rockers and applied radius bars to the truck itself. Spring tension was also adjusted to discourage lateral movement. P5a's equipped with these improvements had performed well at Claymont. At 70 miles per hour, the locomotives made an average impression (according to the Brinnell tests) of 0.027 inch. At 98 miles per hour, they registered an impression of 0.055 inch. Factory versions of the P5a, that is, ones which had not been altered in any way, produced an average impression of 0.036 inch at 70 miles per hour and 0.055 at 78 miles per hour.[55]

The answer to the axle cracks involved a less complicated procedure. The shop forces at Wilmington simply replaced the P5a's original driving axles with ones of larger diameter.

All the P5a's had returned to service by mid-December. The alterations to their design proved successful. Lateral motion had been brought within acceptable limits, and no fissures appeared in the new axles. Unfortunately, no degree of mechanical modification could compensate for the locomotives' inability to handle the heaviest trains at the highest speeds, the most serious of the P5a's weaknesses.[56] Thus five years after it had launched long-distance electrification, the Pennsylvania found itself in a rather incongruous predicament. It had been able to accomplish what many observers had deemed impossible: the successful financing and construction of a vast program of capital improvements during a time of unprecedented national economic distress. On the other hand, the railroad had not achieved a goal which in 1928 it had taken for granted. The PRR had failed to develop a locomotive which would allow it to realize the full potential for efficient operation inherent in the use of electric traction.

7

··

Washington, Harrisburg, and the War

The frustration which the Atterbury administration experienced in trying to obtain a satisfactory class of electric motive power did not cause it to diminish its efforts to extend catenary to Washington. The reverse occurred. In the face of serious financial hardship, the Pennsylvania over the next few years enlarged its electrified territory beyond even what Atterbury had contemplated in his February 1931 announcement. During this same period, the railroad at last solved its locomotive problems. From the P5a debacle emerged what has become generally regarded as one of the finest land transportation tools modern man has yet produced.

Late in 1933, the PRR decided that it had no alternative but to formulate plans for a new locomotive to replace the P5a as the road's standard electric passenger power. The Pennsylvania instructed General Electric and Westinghouse each to build a single prototype of the new engine. The railroad asked that the new machines be more powerful than the P5a, have a lighter axle loading, and be capable of speeds of 100 miles per hour or more. The locomotives were also to have streamlined superstructures with a double-ended cab positioned in the center.[1] The 1930s marked the introduction of flowing lines and graceful contours in industrial design. Automobiles and railroad rolling stock in particular reflected the new trend toward combining an eye-pleasing form with a utilitarian function. The P5a's, with their boxy carbodies, looked increasingly awkward on the head end of the PRR's fastest passenger trains. Furthermore, the box cab design was unsafe. It placed the engineman and fireman at the very front of the locomotive, leaving them virtually unprotected in the event of collision. The consequences could be tragic, as an accident occurring at a grade crossing near Deans, New Jersey, demonstrated. On January 3, 1934, P5a No. 4772, pulling the westbound Spirit of St. Louis at 50 miles per hour, slammed into a truck which had failed to stop at the crossing. The engineman was pinned

P5a No. 4709 at Washington, D.C., 1938. (Harold K. Vollrath Collection)

in the cab and killed. The fireman escaped a similar fate only because he had been attending to the train-heating boiler in the rear of the locomotive at the time of impact.[2]

General Electric proposed to fashion its new electric for the PRR after the ten 3,000-horsepower, Class EP-3 locomotives which it had built for the New Haven in 1931. Employing a 2–C+C–2 wheel arrangement, these articulated engines had no difficulty in maintaining the fast passenger schedules between New Haven and New York. Their duties in that service closely resembled those required of Pennsylvania passenger locomotives. The EP-3's performed so well that the PRR borrowed one (No. 0354) in 1933 to test at Claymont. The PRR wished to compare the lateral thrusts of the EP-3 with those produced by the P5a. Measurements revealed that the New Haven unit showed far less inclination to oscillate than the P5a. At 100 miles per hour, for example, the EP-3 registered an average indentation of only 0.034 inch, as computed by the Brinnell hardness method. This figure compared favorably to the 0.055 inch recorded by an improved P5a at 98 miles per hour.[3]

General Electric recommended a locomotive similar to the EP-3 to the Pennsylvania. The PRR's motive power officials traditionally had little regard for articulated locomotives, of course, either steam or electric. The articulated DD1's had proven extremely capable, but the railroad in 1934 was still disposed to pattern its electric locomotives after its own steam types. The Pennsylvania had never possessed a successful articulated steam locomotive. However, J. V. B. Duer and his associates could hardly ignore the flawless performance of the EP-3. Moreover, George Gibbs, the PRR's consultant on electrification, agreed with GE's engineers. In fact, it was at Gibbs's suggestion that the Pennsylvania had borrowed the EP-3. He had long held the view that to minimize side blows to the rails, electric locomotives should have a relatively light axle loading—certainly no more than 50,000 pounds. The P5a carried a 75,000-pound axle loading. Gibbs and General Electric contended that any locomotive that carried such an excessive rating would always be a source of severe lateral force. True, when the PRR had refined the P5a's suspension system that force had been reduced; but it was still greater than the railroad considered ideal for prolonged, high-speed operation. Lower axle loadings naturally meant lower horsepower per axle. Since the Pennsylvania demanded a very powerful engine, GE's engineers argued, so many driving axles would have to be used that articulation was imperative. Otherwise the rigid frame would be much too long. Other than to stipulate a few basic requirements, the PRR had given the manufacturers a free hand in constructing their test models. During the spring and summer of 1934, therefore, a 4,620-horsepower, 2–C+C–2 locomotive began to take shape at GE's Erie works.[4]

Westinghouse concurred in the opinion that lower axle loadings were necessary to inhibit lateral motion. However, the engineers at East Pittsburgh did not believe that articulation would be of material benefit. It might even be counterproductive. The decision to forgo the use of an articulated frame meant that Westinghouse could not incorporate as many driving axles into its design as GE had, resulting in somewhat higher axle loadings. What began to materialize at East Pittsburgh was a 5,000-horsepower, 2–D–2 locomotive that was little more than an elongated and more powerful version of the P5a.[5]

Westinghouse and General Electric delivered their creations to the Pennsylvania in August 1934. The railroad designated the Westinghouse locomotive Class R1 and gave it road number 4800. Twin motors totaling 1,250 horsepower drove each of the four main axles through geared quill drives. Both wheels of each axle received power. Westinghouse had wisely decided to discard the cheaper single-ended drive after its limitations became apparent in the P5a. Axle loading was rated at 60,000 pounds. General Electric's locomotive was denoted Class GG1 and received road number 4899. It, too, utilized twin motors (totaling 770 horsepower) to power each of the six driving axles through the same kind of geared quill drive found in the R1. The GG1's axle loading was 50,000 pounds. Baldwin supplied the carbodies and some of the mechanical parts for both units.[6] The choice of roster numbers for the new engines betrayed the railroad's suspicion of locomotives having articulated running gear and a large number of axles. If the R1 were to be the PRR's ultimate selection, production models could be numbered consecutively beginning with No. 4801. But if the road favored the GG1 instead, renumbering was inevitable. The Pennsylvania followed a far more orderly procedure of assigning numbers to its electric locomotives than it had to its steam fleet; hence No. 4899 would not be a suitable designation for the first in a series of locomotives of the same class. Even if the PRR were to order both the R1 and GG1 in quantity, the original GG1's number would almost certainly have to be changed, perhaps to No. 4850, for instance. The choice of numbers for the two prototypes hardly represented a PRR vote of no confidence in the GE product. It was simply a subtle, possibly even an unconscious, indication that the railroad's engineers were more optimistic of the R1's chance for success.

At any rate, Duer, Chief of Motive Power Fred W. Hankins (who had succeeded James Wallis in that post in 1927, and their staffs intended to subject both locomotives to the most exhaustive series of motive power tests yet conducted by the Pennsylvania. If either engine harbored any flaws, they would soon become apparent. The railroad had traditionally taken a very cautious and methodical approach when developing new motive power, whether steam or electric. It usually thoroughly tested a few experimental models before making a decision regarding their mass production.[7] The P5a represented an aberration from this practice. The PRR had wanted a full complement of heavy passenger engines on the roster when long-

R1 No. 4899 (formerly No. 4800) at Wilmington, 1937. (William E. Grant)

distance service commenced early in 1933. In its haste to reach this goal, the road had temporarily floundered. This would not be allowed to happen with the R1 and GG1.

As soon as they arrived on the PRR, the new locomotives began ten weeks of rigorous testing, not only at Claymont but also in regular New York-Philadelphia service. Both machines proved that they could easily wheel the heaviest of the railroad's passenger trains at speeds in excess of 90 miles per hour. Both developed short-term ratings of approximately 10,000 horsepower, and both accelerated their trains with remarkable swiftness. Results of track tests at Claymont showed the R1 and GG1 to be superior even to the P5a's which had improved suspensions. According to the Brinnell hardness tests, the R1 and P5a averaged about the same lateral thrust (0.033 inch for the P5a and 0.030 inch for the R1) at 80 miles per hour. At 100 miles per hour, however, the R1 recorded an average impression of 0.046 inch, compared to the 0.055 inch indentation made by the P5a at 98 miles per hour. While the R1 did well, the GG1 did much better. It registered impressions of 0.027 inch at 80 miles per hour and 0.033 inch at 100 miles per hour. Readings taken from the electronic strain gauges confirmed these findings.[8]

The factors favoring the selection of the GG1 over the R1 were not confined to the outcome of the Claymont trials. The railroad had some doubt concerning the ability of the R1, with its long, rigid wheelbase, to safely negotiate sharp curves and yard turnouts. In addition, the GG1's Type 627 traction motors were very similar in size and design to the motors which the Pennsylvania had been using for many years in its fleet of MP-54's. The road's maintenance forces had acquired plenty of experience in their operation and repair.[9]

Duer and Hankins needed no further evidence to convince them that the GG1 would admirably fulfill their railroad's requirements. On November 17, 1934, the Pennsylvania ordered 57 more of the articulated behemoths. Although the locomotive was largely a GE product, the PRR's traditional supplier, Westinghouse, received an order for electrical gear and controls for 34 of the new engines. General Electric provided these items for the remaining 23. General Electric was also to build the carbodies for 14, with Baldwin supplying another 25, and the railroad itself 18. Electrical equipment on all of the locomotives except the 14 constructed at Erie was to be installed at Altoona. Each machine carried a price tag of $250,000 —roughly double the cost of an M1, the Pennsylvania's newest and most powerful steam locomotive.[10]

Soon after it placed the order, the railroad engaged Raymond Loewy, a young industrial designer, to enhance the GG1's aesthetics. Loewy, who would one day become an internationally respected figure in his profession, had just completed his first assignment for the

Builder photograph of GG1 No. 4918. (Altoona Public Library)

PRR: redesigning the trash cans in New York's Pennsylvania Station. Working on consultation with Chief of Motive Power Hankins, Loewy substituted a sleek, all-welded carbody for the riveted shell found on the first GG1. This innovation, by eliminating protruding rivets and overlapping seams, increased the locomotives' aerodynamic soundness. A full-scale mock-up of a GG1 shell was constructed at the Wilmington shops, where Loewy added a few other cosmetic refinements.[11] All new GG1's were to be painted brunswick green (the PRR's previous electrics had been black) with five gold pinstripes along the flank. This scheme was so attractive that it established a pattern which the railroad eventually applied to nearly all its electric (and later diesel) passenger engines for the next 20 years.[12]

When it had opted for a new passenger locomotive early in 1934, the Pennsylvania had no intention of writing off its P5a fleet as an ill-advised and wasteful investment. Rather it intended to use the P5a's to form the core of its electrified freight service. In July 1934, the PRR ordered the remaining 28 P5a's which it had originally contracted for three years earlier. To ensure crew safety, the railroad specified that they be delivered with the same kind of center-cab streamlined styling found on the GG1 and the R1. Internally these "P5a modifieds," as the Pennsylvania termed them, were identical to the other P5a's. The railroad also was preparing to regear the box cabs for lower-speed freight duties. The new modifieds would retain their 90-mile-per-hour gearing in order to augment the GG1's in passenger service if the need arose.[13] The timing of this shift in motive power assignments was extremely fortuitous for the PRR. It had discovered that the electrics initially built for freight service, the L6's, were not sufficiently powerful when used singly to haul heavy freight trains. Had the P5a's proved successful in passenger service, in all likelihood the railroad would have been forced to develop a new freight engine. As events unfolded, the Pennsylvania transferred the P5a's to freight duties and canceled its order for 30 additional L6's. Since Lima had already finished the carbodies by this time, the railroad had no choice but to take delivery on them. Only 1 (Class L6a) ever received electrical components, however. The remaining 29 were stored at Hollidaysburg and scrapped a few years later.[14]

The builders promised that the first of the new GG1's would arrive as early as April 1935. This could not be too soon in the railroad's estimation. Passenger traffic in the New York-Washington corridor had been steadily rising since early 1933. A. H. Shaw, general passenger agent for the New York zone, attributed the increase partly to the improved service wrought by electrification and partly to the flurry of activity in the nation's capital resulting from President Roosevelt's New Deal programs. In any case, the Pennsylvania was regularly adding extra cars to its passenger trains.[15] More powerful locomotives were urgently needed to cope with this most welcome upsurge in business, especially with electrified service to and from Washington's Union Station scheduled to begin early in 1935.

For a time, the railroad had been unsure whether electrification to Washington would be completed at all, at least within the foreseeable future. During the last half of 1933, as the PRR struggled to cure the ills of its P5a's, the financial reserves which had been sustaining the advance of electrification had dwindled almost to the point of nonexistence.

The Pennsylvania had never been entirely satisfied with the terms of its loan from the Reconstruction Finance Corporation. The road's directors considered the 6% interest rate to be unjustifiably high for so large a loan. The executive committee of the RFC expressed a willingness to lower the rate, provided that the railroad met certain conditions. First the committee demanded that the railroad submit a list of salaries received by its officers and directors. If the RFC found these salaries to be "reasonable," or if the PRR would subsequently

P5a modified No. 4787 on the point of a New York-Philadelphia "clocker" near Trenton, New Jersey in 1938. (Harold K. Vollrath Collection)

reduce any "unreasonably high" salaries, the government agency would grant the Pennsylvania a 5% interest rate. President Atterbury, backed by his company's board, refused to furnish the desired list, even though all of the railroad's top officials had taken substantial pay cuts every year since 1929. Instead the PRR decided to return forthwith the entire $27 million which it had borrowed from the government. On June 30, 1933, it made an initial repayment of $5 million. On July 3, the RFC's executive committee voted to offer the railroad a 5½% rate of interest on the debt still outstanding. If the government hoped that this action would persuade the Atterbury administration to retain the remainder of the loan, it was to be disappointed. The Pennsylvania continued to pay back the money, and on July 28, 1933, it made a final outlay of $6.4 million. The PRR thereby became the first railroad to repay in full a major RFC loan. (In 1933, railroads ranked second only to financial institutions as the RFC's largest borrowers.)[16]

The absence of RFC funds left practically no money for the railroad to use in continuing electrification work, other than to try to correct the deficiencies of the P5a. An attempt to find buyers for a bond issue met with no more success than had the previous effort a year and a half earlier. By summer's end, construction on the partially electrified section of main line south of Wilmington ceased.[17]

Work came to a halt at a time when the federal government was launching a multitude of new bureaus whose long-range objective was to restore the nation's economic health. Among the agencies which the Roosevelt administration charged with promoting employment and reviving the sagging demand for industrial goods was the Public Works Administration (PWA), created by Congress at the President's urging in June 1933. The PWA's primary task was to help finance the construction of large capital projects such as dams, bridges, ships, and public buildings. To head the PWA Roosevelt chose Harold L. Ickes, who served concurrently as secretary of the interior. Ickes's deputy at PWA, an army engineer named Colonel Henry M. Waite, had been closely following the course of PRR electrification. When the railroad found itself unable to support the project any longer, Waite suggested to his superior that electrification would make a fine target for PWA funding. Ickes agreed. On August 20, Ickes and Waite discussed the proposal with President Franklin D. Roosevelt, who also liked it and directed the two men to explore the matter further.[18]

Ickes dispatched Waite to Philadelphia to lay the proposition before Atterbury. The Pennsylvania's president seemed very cool to the idea of taking out another government loan. Ickes then sent Frank Wright, head of the PWA's Transportation division, to confer with Atterbury and other influential PRR officials. At length the railroad agreed to request a loan of some $80 million. Ickes made sure that the application was rushed through the proper channels at top speed, especially at the Interstate Commerce Commission, an agency notorious for its bureaucratic sluggishness. On November 2, 1933, "a big day in public works," according to Ickes, the PWA formally approved a loan of $84 million to the railroad.[19] This was the largest loan granted by the PWA to that date, and the first loan it had made to a railroad. Ickes was delighted. He expected (correctly) that the precedent set by the Pennsylvania would convince other needy lines to make similar requests for funds. Thousands of jobs would ultimately be created. Before the PRR's representatives would sign the loan agreement, however, they desired to consult further with PWA officials. In negotiations conducted over the next several weeks, the railroad scaled down its request to $77 million. It wanted to borrow not a penny more than was absolutely necessary. Atterbury also convinced Ickes to proclaim a one-year moratorium on collecting the 4% interest the loan carried. The two parties finally signed the agreement on December 29.[20]

The Pennsylvania planned to spend approximately $45 million of the loan to complete the electrification to Washington and $15 million for new electric locomotives. The remaining funds were to be divided between electrifying a number of freight yards and building 7,000 new freight cars. In return the PRR pledged to abide by certain PWA regulations. Unskilled laborers were to receive a minimum wage of $15 per week. All materials were to be purchased whenever possible from manufacturers who subscribed to the code of the National Recovery Administration. The railroad was to provide the PWA with monthly progress reports detailing where and how much money was being spent.[21]

President Atterbury predicted that the loan would create the equivalent of one year of work for 25,000 men, including peripheral employment in the railway supply industry. By mid-1934, over 10,000 men labored on the various electrification-related projects. Later computations showed that 43% of the PWA dollars went directly to workers as wages and salaries.[22]

The PWA loan also played an important role in alleviating the PRR's motive power woes. The railroad had resolved to find a new electric passenger locomotive to supersede the P5a at about the same time it had asked to borrow money from Ickes's agency. The Atterbury administration would have been extremely hard pressed to carry its search for a satisfactory locomotive to a successful conclusion had it not received support from the Public Works Administration. Prior to the granting of the PWA loan, GE and Westinghouse each had spent $350,000 on research and development work aimed at finding a replacement for the P5a; but the railroad could not afford to pay the manufacturers even that amount until its depleted treasury had been invigorated by an infusion of PWA dollars.[23]

Although President Atterbury's statement of November 1, 1928, contained no reference to the possible extension of catenary beyond Wilmington, the Pennsylvania considered electrification of the full length of the New York-Washington corridor to be mandatory if all the economies of electric operation were to be realized. Atterbury did not include the Wilmington-Washington segment in the road's immediate plans principally because an unexpected obstacle had arisen at Baltimore. Municipal officials there had long been pressuring the PRR to electrify operations through the city as a means of combating the terrible smoke problem. However, Baltimore's city fathers would not permit the Pennsylvania to run 132,000-volt transmission lines through the city unless the lines were buried. The railroad balked at this demand. It protested that it had no experience with the construction and maintenance of high-voltage subterranean cables. (Lines energized at only 11,000 volts ran beneath the Hudson and East rivers.) The two sides wrangled over this question for many months. In 1930, the PRR finally conceded the issue. It was anxious to advance its improvement program while labor and material costs were still unusually low.[24] The following year, Atterbury publicly announced his company's intention to convert its New York-Washington main line to electric traction.

This decision made urgent a number of related improvements in the Baltimore area that the Pennsylvania had been studying for several years. The old 3,450-foot Union Railroad tunnel east of the main passenger station could not accommodate two tracks for electrified operation owing to clearance problems with the overhead wires. The railroad therefore removed one of the tunnel's tracks and shifted the other to the center of the bore, where proper clearance could be obtained. It also constructed a new double-track tunnel adjacent to and parallel with the Union. The PRR planned to take similar steps to relieve congestion through the antiquated Baltimore and Potomac tunnel (built in 1872) which lay southwest of the station. Unfortunately, even with PWA money, the railroad did not possess sufficient funds to undertake this project. Instead it relined the old tunnel with concrete (to prevent dripping water from freezing and interfering with overhead current collection) and abandoned plans for a new tunnel. The PRR did have enough cash to eliminate some hazardous grade crossings in the city, as well as to add more running tracks near the station and to lengthen station platforms. The longer train lengths made possible by electrification necessitated longer platforms.[25]

The Pennsylvania required an act of Congress—which it secured in routine fashion—to allow it to string high-voltage lines in the District of Columbia. While electrified passenger service terminated at Union Station, the railroad continued to run wires through the Virginia Avenue tunnel and across the stone bridge spanning the Potomac River to Potomac yard, near Alexandria, Virginia. This yard was the PRR's principal gateway for freight traffic to and from the south. The tunnel presented the usual clearance restrictions. One of its two tracks had to be removed and the other lowered 2 feet before the passage could accommodate electric trains. The river crossing demanded the use of submarine transmission cables.[26]

The PRR experienced no difficulty in obtaining an adequate supply of electric power for that portion of the main line lying beyond the service area of Philadelphia Electric. As early as 1928, the Atterbury administration negotiated with Consolidated Gas, Electric Light, and Power Company of Baltimore with a view toward having the utility act as sole supplier for the southern end of the electrification. Consolidated Gas welcomed the railroad's business. The additional load, it claimed, would reduce the unit cost of power generation, thus benefiting all of the utility's customers.[27] In October

1931, the PRR and Consolidated Gas signed a long-term contract. The utility and two of its affiliates, the Pennsylvania Water and Power Company and the Safe Harbor Water Power Corporation, pledged to furnish electricity to the railroad for a 20-year period beginning in 1933. Some of the current was to be generated by Consolidated Gas's own coal-fired plants near Baltimore. Most was to come from the new hydroelectric station at Safe Harbor, Pennsylvania, 30 miles upriver from Conowingo. One generator at Safe Harbor produced 25-cycle, single-phase power. The other generators there and at Baltimore produced 60-cycle, three-phase power. The railroad thus had to have rotary converters installed at Safe Harbor and at Benning (in the District of Columbia) to adapt this current for rail use.[28]

Responsibility for the maintenance of an adequate flow of electricity from the utilities to the railroad belonged to the load dispatcher, whose office was located in the new Thirtieth Street Station in Philadelphia. In cooperation with several power directors scattered throughout electrified territory, the dispatcher also made sure that the amount of current available at any given time and place did not fall short of the demand for it.[29]

Work on the electrification to Washington and related improvements resumed its previous intensity by the summer of 1934. The Pennsylvania now had enough money to finance the project, and it would soon have a locomotive (the GG1) capable of meeting practically any operating challenge. With the end of the great enterprise at last in sight, President Atterbury relinquished his leadership duties long enough to undergo a brief hospital stay. (He suffered from chronic gallstones.) In July 1934, he underwent abdominal surgery, an operation from which he never fully recovered. Realizing that his failing health prevented him from resuming his responsibilities, Atterbury declined reelection to his company's presidency in April of the following year. He died five months later. His death was the second blow within two years inflicted upon the railroad's executive ranks. Vice-President Elisha Lee had succumbed to a heart attack in 1933. In Atterbury's retirement message, he had hailed the PRR's new vice-president, Martin W. Clement, as "unquestionably the ablest railroad executive in the country" and had urged the directors to name Clement the company's new chief executive.[30] The board complied. It probably would have selected the 52-year-old Clement even without such a glowing endorsement.

A native of Sunbury, Pennsylvania, Martin Withington Clement had entered the PRR's employ in 1901. He worked his way upward through the engineering ranks before transferring to the operating department, where he held several important managerial posts beginning in 1917. At the outset of electrification in 1928, Clement occupied the road's number three position, vice-president in charge of operations. He advanced to the office of vice-president upon Lee's death. Throughout these last few years, while Atterbury and Lee looked after the planning and financing of electrification, Clement supervised the day-to-day tasks of actually implementing the project. He was widely regarded as a man who had intimate knowledge of all facets of his company's operation.[31]

Clement had already assumed the informal role of acting president by the time the Pennsylvania prepared to operate trains electrically the entire distance between the nation's capital and New York. Before commencing regularly scheduled service, however, the railroad decided to mark the occasion by running a special train from Washington to Philadelphia and return. On January 28, 1935, Vice-President Clement, Secretary Ickes, and over 100 other railroad, government, and civic notables assembled at Washington's Union Station to board the special flyer, which consisted of GG1 No. 4800 (formerly No. 4899) and nine cars.[32]

Prior to embarking on their journey, Clement and Ickes made the appropriate dedicatory remarks, which local radio stations aired live. Ickes stepped to the microphone first. He spoke almost as if Franklin Roosevelt deserved most of the credit for successfully completing the improvements. After thanking the railroad for cooperating with the federal government "to the fullest extent and in every possible way," Ickes said the electrification "shows not only what can and should be done under the President's recovery program, but demonstrates what actually has been accomplished under PWA when private initiative aids the Administration in carrying out its re-employment plans."[33]

If Clement believed that in reality it was the administration which had aided private initiative, he did not appear perturbed by Ickes's emphasis on the role of government. He thanked Ickes for his personal assistance and then acknowledged the importance of federal help.

> We are happy to have this opportunity to express our appreciation to the Administration for the pleasure we have had in working with them, and in their having made the completion of this project possible. . . . In a year's lapse of time, through perfect cooperation without any friction, a smooth-running army of railroad employees numbering in the thousands, backed by many more in industry, have again come through in the American way—on time.[34]

The conversion of the New York-Washington main line to electric traction, concluded Clement, "represents the faith of this railroad not only in rail transportation, but in the future of our country."[35]

The ceremonies over, Clement, Ickes, and their guests boarded the cars for the trip to Philadelphia. The railroad made no attempt to set a new speed record. Except for the free champagne, the run was designed to approximate normal operating conditions as closely as possible. Nevertheless, GG1 No. 4800 and its train performed exceptionally well. Total time clocked for the round trip was only 3 hours, 54 minutes. The special posted an average speed of 74 miles per hour, although between Landover and Seabrook, Maryland, it attained a speed of 102 miles per hour (on the outbound leg of the trip). On the return, the train established a new speed record in spite of the railroad's indifference, arriving at Washington just 1 hour and 50 minutes after leaving Philadelphia.[36]

Electric service for fare-paying passengers began on February 10, 1935. By April 8, all 191 through passenger trains between New York, Philadelphia, and Washington were being hauled by electric locomotives. In addition, 448 m.u. trains operated in local service.[37] In April, freshly outshopped GG1's started supplanting the P5a's which had until now been pulling the trains. The arrival of these new locomotives allowed the Pennsylvania to electrify most of the freight trains which it operated on the New York-Washington route. The railroad had always considered electrification of freight service in this corridor to be of secondary importance. Passenger operations netted it less revenue—even in this densely populated region—than did freight. Yet the passenger business was far more vulnerable to the threat of competition. Still, the Pennsylvania was not about to overlook the economies which electric traction offered for freight trains. In a press release issued in January 1935, the railroad noted that

> freight traffic is of vital character, including as it does enormous quantities of fresh fruits and vegetables from the South to the great markets north of the Potomac River and east of the Allegheny mountains. Prompt delivery of this freight is essential to meet the requirements and customs of the trade. The greater speeds made possible by electric operation will not only permit quicker schedules but insure a much higher degree of dependability of arrival.[38]

Through freights ran behind electric locomotives only as far south as Bay View yard, Baltimore, until mid-June, when wires leading to Potomac yard were finally energized.[39]

Electrification began paying dividends well before the PRR completed the final link to Washington. By using electric locomotives to haul its east-west passenger trains between Paoli and New York, the railroad was able to significantly reduce running times. The Pennsylvania Limited, for example, beginning in April 1934 required just 20 hours to travel from New York to Chicago. This bettered its previous schedule by by nearly 2 hours."[40] A year later, the Pennsylvania allotted the Broadway Limited 17 hours to make the run between New York and Chicago, bettering the former time by 45 minutes. Effective September 29, 1935, the PRR decreased the timetable by an additional half hour, so that the Broadway's new schedule matched that of the New York Central's Twentieth Century Limited. Also beginning on that date, the Congressional Limited regularly made the run from New York to Washington in 3 hours, 45 minutes, a half hour faster than a year earlier. The schedules of 44 other New York-Washington trains were similarly speeded up. The Pennsylvania cut the running times of its New York-Philadelphia "clockers," too, the best time being 1 hour, 40 minutes. The railroad continued to accelerate these schedules into 1936. By the end of that year, the Congressional required a mere 3 hours, 35 minutes to reach Washington from New York. When still steam powered, the Congressional's fastest regularly carded time had been 4 hours, 15 minutes.[41]

The railroad successfully implemented the new schedules—which affected virtually every passenger train that was electrically propelled—in spite of its simultaneous introduction of air conditioning to its through passenger car fleet. (The Pullman Company also began to air-condition those of its cars which the PRR normally operated.) The air-conditioning machinery materially increased the weight of each car. By August 1933, every passenger car in through

GG1 No. 4829 leading an all-coach test train between Philadelphia and Paoli, about 1937. (Altoona Public Library)

service in the electrified corridor had been air-conditioned. A year later, the Pennsylvania was able to boast that it owned the largest fleet of air-conditioned cars in the world. To be sure, electric traction alone was not responsible for the increase in train speeds, particularly on east-west runs. The railroad started applying mechanical stokers in the early 1930s to the steam locomotives which handled its long-distance passenger trains. The stokers brought about a noticeable increase in the efficiency of these engines. At the same time, the Pennsylvania began experimenting with larger tenders for some of these locomotives, permitting the steamers to make fewer stops for coal and water.[42]

Improvements in trains speeds enabled the PRR to establish more convenient connections with trains from the south. The railroad began to provide third-morning rather than fourth-morning delivery of Florida perishables to the New York market, for example. The fastest freights now made the run between Potomac yard and New York in 6½ hours, eclipsing the best time of steam-powered freights by 2½ hours.[43]

Electric traction contributed to quickened schedules primarily by allowing higher average speeds, as opposed to higher maximum speeds. Because of their locomotives' ability to absorb an overload for short periods, electric trains ascended grades with little or no loss in speed. They accelerated much faster than steam-powered trains after station stops.

The improved passenger service in electrified territory attracted more riders. For the period April 1 to December 31, 1935, the PRR's passenger revenues from New York-Washington trains climbed 5.6% from the same period the previous year. Passenger revenues for the system as a whole were up only 2.2%. C. H. Mathews, Jr., general manager of passenger traffic, estimated that electrification accounted for 3.0% of the difference in the two rates. The remaining 0.4%, he judged, was derived from an increase in Florida traffic. President Clement was eager to make a more precise evaluation of how electrification had influenced both passenger and freight revenues and of how much electric traction had saved the railroad in operating expenses. The road's new vice-president of operations, J. F. Deasey, attempted a preliminary examination of these questions during the fall of 1936 but soon reported to Clement that not enough time had elapsed since the commencement of electric operation. Reliable figures would not be available for another year or two.[44]

In his 1929 statement of the Pennsylvania's electrification plans, Atterbury had indicated that the railroad expected to electrify the Low Grade Line and the Trenton Cut-Off in addition to the New York-Washington main line. As of 1936, however, the freight routes still lacked catenary. The PRR had devoted its scarce financial resources to its more heavily traveled north-south corridor. Clement and Deasey realized, nonetheless, that even in the absence of extensive data confirming electrification's benefits thus far, the railroad would soon have to push the limits of electrified territory farther westward. Freight trains bound to and from the west rolled under wires only for the 50 or so miles between Morrisville and the north Jersey yards. This distance was not of sufficient length to justify the use of electric locomotives, and most east-west freights continued to be powered by steam. The longer (111 miles) run between Paoli and New York made by passenger trains did warrant the use of electric traction, but even there advantages were slight in comparison to the size of capital investment. Such short hauls resulted in poor utilization of motive power. Locomotives spent too much time standing idle at either end of the line.[45]

In January 1937, the PRR's directors, heeding President Clement's advice, voted to electrify both freight and passenger main lines west to the Susquehanna River. They also authorized the electrification of the passenger main line from Paoli to Harrisburg. This measure would not only provide a longer electrified run for New York-bound trains, it would permit electric locomotives to haul the considerable number of east-west trains that still originated or terminated at Philadelphia's Broad Street Station. That decrepit structure had not been razed following the opening of the PRR's two new Philadelphia stations, chiefly because the city could not contribute its share of funds to the planned redevelopment of the area. New York-Philadelphia "clockers" and a few other long-distance and semilocal trains still used the old terminal. Another element figuring into the continued utilization of Broad Street Station was the inability of the GG1's to enter Suburban Station, owing to clearance restrictions. If the Pennsylvania were to abandon Broad Street, it would no longer have a downtown terminal for its New York-Philadelphia trains (which, of course, were powered by GG1's). It would then risk losing some of this business to the Reading, which did maintain a downtown (Market Street) terminal.[46]

Other trackage scheduled for electrification included the Columbia and Port Deposit Branch and a line linking Perth Amboy, New Jersey, with the main line at Monmouth Junction. The "Port Road" ran from a junction with the Low Grade Line near Columbia, Pennsylvania, south along the Susquehanna to Perryville, Maryland. It served primarily as a conduit for freight traffic bound to and from Baltimore and points south. Electrification of the Perth Amboy line facilitated movements between the north Jersey terminals and the main line. The railroad also slated Enola yard for electrification and expansion. Already a sprawling freight terminal, Enola was to become the key interchange point for steam and electric motive power. The PRR planned to electrify a total of 315 route-miles and 773 track-miles. It would then have 2,677 track-miles under electric operation, or about 41% of the total electrified track-mileage in the United States.[47]

A business upturn throughout most of the nation in 1936 and 1937 created a ready market for the $52 million worth of bonds which the railroad issued to finance the venture. Construction began in the spring of 1937, with completion scheduled for the fall of the following year. Over 10,000 men labored on the project at the peak of activity. In contrast to the PRR's reliance on the labor of private contractors, a large portion of this work force was on the railroad's own payroll. When the Pennsylvania had resumed work on electrification in 1934 after receiving PWA money, it called back a number of furloughed employees (whose former occupations ranged from clerk to engineman) and trained them for such new positions as electrician, linesman, and carpenter. While many experienced electrical construction workers were available, the railroad believed it had a duty to hire the men whom it previously had laid off. This was also a good means by which to gather a permanent, experienced group of workers whose task would be to repair and maintain the electrification once construction was finished. The Harrisburg electrification reflected an expansion of this practice. The PRR recalled even more former employees, many of whom then worked temporarily under the supervision of private contractors until the project's completion.[48]

Probably this generous labor policy undercut employee resistance to electrification to some degree, since many men who had lost their old jobs because of electric traction's labor-saving characteristics were subsequently rehired as linesmen and electricians. Another factor which helped to neutralize what little animosity organized labor held toward electrification was the depression itself. With the railroad furloughing workers by the thousands, the number of men thrown out of work by the introduction of electric traction must have seemed minuscule. And insofar as electrification promised to strengthen the PRR in its battle with competing modes of transport, it obviously worked to the employees' long-term advantage.

Construction proceeded smoothly throughout 1937. Since much of the route to be electrified consisted of single and double track, the Pennsylvania was able to utilize the cheaper, less intricate catenary structures. The road also placed its communications and signal lines in armored cables suspended from poles along the right-of-way, a less expensive and time-consuming method of shielding the lines than the earlier practice of burying them. The Philadelphia Electric Company readily consented to supply power as far west as Lancas-

Workers erecting catenary between Paoli and Harrisburg, 1937. (PRR)

ter. The railroad contracted with the Safe Harbor Water Power Corporation to furnish current beyond that point. The utility augmented the single-phase generating capacity of its Safe Harbor hydroelectric station to meet the anticipated increase in load.[49]

The rapid pace of construction surpassed the Pennsylvania's expectations. Less than a year after work began, the first regularly scheduled electric passenger train made its way under the wires from Philadelphia to Harrisburg. The Metropolitan, consisting of GG1 No. 4863 and 13 cars, arrived at Harrisburg Station from Philadelphia at 12:08 P.M., January 15, 1938. The shrieking whistles of countless PRR locomotives, mingled with the blasts of assorted factory whistles, fire sirens, and automobile horns, saluted the train along the length of its journey. As the Metropolitan sped through West Philadelphia, it evoked such an ear-rending clamor that many unsuspecting citizens reportedly believed Nazi Germany had just invaded. Others supposed that a huge fire was sweeping the city. A

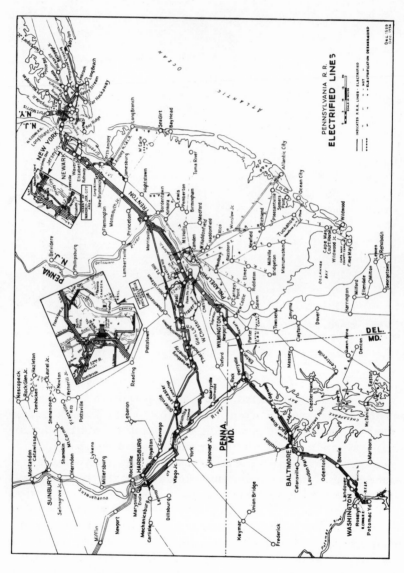

Pennsylvania Railroad electrified territory. (PRR)

reporter for the *Philadelphia Inquirer* declared that the cacophony "sounded like all the whistles in the world—like New Year's Eve, the first Armistice, and the crack of doom all rolled into one."[50] At Harrisburg, a tremendous crowd gathered to greet the train's arrival. Once the GG1 had brought its consist to a halt, thousands of people swarmed about the locomotive and cars, taking pictures, listening to the dedicatory remarks, even besieging the train's crew for autographs.[51]

Within the next few weeks, electric traction had displaced steam at the head of nearly all PRR through trains east of Harrisburg. By mid-April electrification of the Low Grade and Columbia and Port Deposit routes was completed, and P5a's began handling freight assignments on these lines.[52]

Once the extension to Harrisburg became fully operational, the Pennsylvania undertook a preliminary investigation of electrification's results. The main goal of the study, as determined by the PRR's engineers and representatives from Gibbs and Hill, was to ascertain to what extent electric traction had reduced operating expenses.[53] The findings, embodied in a series of reports issued in 1938–39, confirmed the efficiency of electric motive power. Subsequent studies conducted over the next several years corroborated these initial assessments.

The railroad and its consultants reckoned that electrification lowered the Pennsylvania's motive power operating costs (including maintenance and repair) by $7,701,360 in 1938. This calculation was remarkably close to the $7,500,000 which the PRR three years earlier had predicted it would save.[54]

The railroad found electrics to be significantly cheaper to maintain and repair. It pegged maintenance and repair costs on a per mile basis at 10.4 cents for the GG1 and 14.8 cents for the P5a. Admittedly, both types were newer than the K4s's, which first appeared in mass quantity in 1917 and which in 1938 cost the PRR 21.5 cents per mile to maintain and repair. Yet the GG1 and the P5a compared even more favorably with the more modern M1 and M1a classes of steam engines. The Pennsylvania acquired 300 of these big 4-8-2's between 1926 and 1930 to serve as fast freight and passenger haulers. Their combined average maintenance and repair cost for 1938 was 27.9 cents per mile. The road's two other steam-powered mainline freight classes, L1s and I1s, averaged 33.3 cents and 36.4 cents, respectively.[55]

Electric locomotives also spent less time at terminals undergoing inspection and repair. The GG1 and P5a classes required approximately one hour per day for routine inspection, a chore which was usually accomplished while the engines were lying over between assignments. Daily inspection periods for the larger steam types ranged from four to five hours. The monthly inspection to which the railroad subjected each locomotive (in accordance with ICC regulations) required one day for electrics and three days for steamers. By 1940, the PRR was performing class (heavy) repairs on its GG1's and P5a's at intervals of 350,000 miles, while executing the same operation on its main-line steam locomotives at 90,000 miles. Because the electric locomotive demanded less attention from shop forces, its record of reliability and availability easily exceeded that of the steam engine. In 1939, PRR electrics (primarily GG1's and P5a's) averaged 7 minutes of detention per 10,000 miles, as contrasted with 25 minutes of detention per 10,000 miles tallied by the steam locomotives they had replaced.[56] In 1940, the combined average annual availability rate of the GG1 and the P5a classes stood at 90% or about 335 days of the year. The several classes of steam locomotives which had preceded the electrics in the same service had shown a combined average annual availability rate of 69%, or 251 days. Even after taking into account the more recent vintage of electric motive power, J. V. B. Duer affirmed that "there is no question that the electric locomotive is a more available unit than the steam locomotive."[57]

The adaptability of electric locomotives to multiple-unit control further enhanced both their availability and their lower operating expense. The railroad could transfer them throughout electrified territory as traffic demanded simply by double- or triple-heading the units on a regular train, using a single engine crew. By the same token, the railroad could add more units to a particularly heavy train without worrying about additional labor costs. Steam locomotives, on the other hand, required a separate engine crew for each machine.[58]

Electrification brought about a significant reduction in fuel costs. The PRR's engineers estimated that a given amount of coal burned in an electric generating plant produced twice the amount of horsepower at the rail if used to create current for electric locomotives than it would if burned in the fireboxes of steam locomotives. In the early 1940s, railroad tests revealed that an L1s pulling a train of 2,470 gross tons between one of the north Jersey terminals and Enola consumed 71.7 cents worth of coal per mile. A P5a hauling the same train drew 56.2 cents worth of electricity per mile—a figure which included the cost of generating the current and then transmitting it to the locomotive. A comparison between the K4s and a GG1 in passenger service revealed a similar ratio of fuel costs. Considering that the Pennsylvania's electric locomotives were accumulating over 30 million miles every year, electrification was saving the railroad a hefty sum on fuel bills alone.[59]

Electrification of the lines east of Enola also released many steam locomotives for service elsewhere on the system. Two hundred seventeen steamers (chiefly of classes K4s and L1s) had been relieved) of their main line assignments after the inauguration of electrified operation between New York and Washington. The Harrisburg extension made surplus an additional 170 steam engines. The Pennsylvania retained many of these locomotives east of Enola to work non-electrified branches and to augment the electric fleet as traffic fluctuations warranted. The rest were dispatched to points west of Enola. In either area, the surplus of newer, more powerful steam engines permitted the railroad to retire older, less efficient types.[60]

The PRR could not determine precisely to what degree electrification had been responsible for the increased business which the railroad enjoyed after 1938, especially since traffic levels were rising nationwide as the country prepared for another war. Still, by all tangible indications the conversion to electric traction was repaying the Pennsylvania handsomely. Senior Electrical Engineer H. C. Griffith aptly summarized his company's feelings.

> The application of electric traction to the eastern territory of the Pennsylvania Railroad . . . has produced a faster, better, and more reliable service with fewer locomotive units and greater economy than is possible with the steam service which it replaced, and has thus placed a more reliable, cleaner, and faster service at the command of the railroad and the traveling public.[61]

To be sure, electrification was peculiarly vulnerable to certain hazards that were of little account in steam operation. One such hazard was malicious mischief. "We have just one missing link in our conquest of the thousand and one problems of electrified operation," remarked a PRR electrical engineer in 1937, "and that is how to teach little boys not to play with high tension electricity."[62] Insulators were a common target for youths with BB guns and .22-caliber rifles. Vandals who lacked more sophisticated weapons often dropped pieces of scrap metal through crevices in overhead bridges to short-circuit the high tension wires below. At least children could be educated to the dangers of electricity—animals could not. The railroad recorded innumerable cases where birds, four-footed creatures, and even swarms of insects found their way into substations (and in several instances, locomotives), causing short circuits in the equipment—and electrocuting themselves in the process, of course. Whenever such events occurred, the railroad frequently had to shut off power to an entire segment of main line, thus delaying all trains in the vicinity.

By far the most troublesome problem was lightning. Although the catenary and substations incorporated many protective features, lightning could still be counted on to produce about half of all power failures in any given year. Fortunately, lightning-caused difficulties were usually the easiest for the PRR's maintenance forces to remedy, with most outages lasting no more than a few minutes.[63] Another meteorological problem—a snow storm—caused the most serious and in fact the only long-term disruption of the Pennsylvania's electrified operations. A fierce blizzard striking the Northeast in February 1958 whipped fine snowflakes through the air filters of the GG1's and short-circuited their traction motors.[64] Aside from this single episode, however, the PRR encountered no problems with electrification that it had not anticipated.

If the conversion to electric traction met with the railroad's enthusiastic approval, it did fail to meet one important expectation of the equipment suppliers. W. D. Bearce, head of General Electric's Transportation department, spoke for his own firm and for Westinghouse when he said in 1935: "The endorsement of railroad electrification by this outstanding railroad system should lead to further applications of electric power for handling mainline freight and passenger, as well as branch-line and switching, traffic."[65]

Prior to 1929, a number of roads had long-distance electrification under close scrutiny. Some, such as the New York Central and the New Haven, were considering enlarging territory already electrified, while others (the Lehigh Valley, for example) were weighing the merits of electric traction for the first time. The shrunken traffic levels and reduced corporate earnings of the 1930s, however, caused most of these railroads to continue with steam. Only two lines undertook significant electrifications during the 1930s, and both ventures involved short-haul traffic. The Reading implemented a 25-cycle, single-phase system utilizing m.u. cars to improve the efficiency of its Philadelphia suburban service. The Delaware, Lackawanna, and Western used a 3,000-volt d.c. system for handling New Jersey commuter traffic and freight switching service in the vicinity of its Hoboken terminal. Insofar as both of these projects had been planned and approved before the onset of the depression, neither constituted an endorsement of the Pennsylvania's electrification.[66]

The PRR electrification soon proved to have a national impact in spite of the industry's failure to follow its lead. World War II presented the fleet of GG1's and P5a's with their finest hour. Wartime freight traffic in electrified territory rose 40% above predepression levels. So many passengers, civilian and military, rode beneath the catenary that the railroad was hard pressed to find cars to accommodate everyone. The central seaboard of the East Coast became the staging area from which the war effort in Europe was mounted. But electric traction so efficiently coped with this crisis that it helped to prevent a recurrence of World War I's transportation nightmares. The PRR seldom found it necessary to reroute trains away from the heavily traveled electrified corridors. No danger existed of coast-bound traffic backing up to Altoona or Pittsburgh. Rarely were trains stranded for long periods (east of Enola) for lack of motive power.[67]

The electrification withstood the terrible ordeal with hardly a sign of strain. By 1943, power consumption had jumped 50% from 1939 (to 150,000 kilowatts per hour, with peaks of 235,000 kilowatts). Yet the utilities were capable of supplying the railroad with 300,000 kilowatts, leaving a comfortable margin for future increases in consumption. The Pennsylvania had but 139 GG1's on its roster at the close of 1943 and had not purchased a P5a since the order for 28 streamlined versions in 1934. Yet the railroad needed relatively few steam locomotives to supplement the electrics. During the fall of 1943, for example, a month that saw a prodigious amount of traffic pass through the electrified region, the PRR had in regular service under the wires less than 100 steam engines.[68]

In full operation for less than a decade, the Pennsylvania's electrification had proven itself to be of inestimable value not only to the railroad but to the nation. More than one knowledgeable observer credited the PRR's conversion to electric traction for having headed off nationalization of the railroads during the war.[69] But even those few short years of peace prior to Pearl Harbor were of sufficient duration to have convinced the Pennsylvania of the wisdom of its decision to electrify. Technologically and economically, electric traction had performed herculean feats up to 1945. Its future seemed bright indeed as the PRR prepared to meet the demands of a nation once again at peace.

8

....

The Challenge of a New Era

In the post-World War II era, two main issues dominated the Pennsylvania Railroad's activities with regard to electric traction. For over a decade following the war, the railroad intermittently considered extending catenary westward from Harrisburg to Pittsburgh, a distance of 245 miles. At the same time, the PRR, with the cooperation of General Electric and Westinghouse, implemented a major electric motive replacement program, a program highlighted by a number of important technological innovations.

The Pennsylvania's financial capability continued to play a prominent role in electrification matters, just as it had in the past. The prosperity that the railroad enjoyed prior to 1929 never returned, a condition which was to influence profoundly the future of electric traction. Nor was the electric locomotive the only challenge to steam during these postwar years. The diesel locomotive finally made its appearance as a high-speed, heavy-duty machine worthy of mainline responsibilities. The diesel offered many of the advantages of the electric but did not carry such a high first cost. Diesels easily banished steam from the PRR and for a time threatened to do the same to electric traction. Yet in the end, the GG1's and their offspring survived both the competition of the diesel and the fiscal adversity, only to see the railroad which they had served so nobly lose its identity and a great deal more in the Penn Central merger of 1968.

The mighty Pennsylvania Railroad had precisely 40 years left in its corporate life when Vice-President Elisha Lee acknowledged that his company expected to electrify its main line the entire distance between New York and Pittsburgh. "Economic conditions will compel us to electrify gradually," Lee cautioned. The main line to Washington demanded top priority. Many years might elapse before the first electrically powered train entered Pittsburgh. Nevertheless, neither he nor Atterbury had much doubt that such an eventuality would come to pass.[1]

In 1930, the PRR disclosed that it was already formulating plans for electrifying the Middle and Pittsburgh divisions. The railroad was then moving 6,250 freight cars daily through the Pittsburgh gateway. Its planners predicted that in five or six years that figure would climb to 8,000 cars and the main line through the mountains would have reached its capacity with steam power. Operational necessity, if not economic choice, would then mandate stringing wires west from Harrisburg. Furthermore, if passenger trains could proceed electrically all the way from New York to Pittsburgh, running times of the New York-Chicago limiteds could be reduced to as little as 14 hours—short enough to better the New York Central's best time by about 2 hours and perhaps undercut competition from the airlines before they could ever become a serious threat.[2]

The Atterbury administration in 1930 had not foreseen the severity or the duration of the Great Depression. Even after the completion of the electrification to Harrisburg, the train-handling capacity of the two divisions to the west had not been met. In 1929, gross ton-miles per route-mile on the Middle division totaled 88,570,000, while the combined total of the Philadelphia division main line and the Low Grade Line stood at 86,830,000. Ten years later, gross ton-miles per route-mile had dropped to 63,250,000 and 47,100,000 on the respective divisions.[3] The four-track line through the rugged Alleghenies remained an enticing target for catenary in spite of the decrease in traffic. Even at low ebb, traffic density was still greater than that found on most other roads in more prosperous times. Because of the steep grades and the PRR practice of running heavier trains west of Harrisburg, double-heading and often triple-heading of motive power were common occurrences. The advent of the more powerful Ml steamers had helped some, but not nearly enough.[4] In short, operating conditions between Harrisburg and Pittsburgh had changed very little from the years before World War I when the PRR had originally considered electrifying this line. Making the prospect of electrified service west of the Susquehanna even more attractive were the achievements wrought by electrification on the New York-Washington line and the continuation of abnormally low construction costs.

The federal government also believed that electrification of this route warranted further study. In the mid-1930s, the Federal Power Commission conducted a survey of the nation's electric power needs and the ways in which electricity could be utilized most effectively in the future. The agency singled out 12,500 route-miles of railroad that it believed could be profitably electrified. Included in this mileage was the PRR's main line from Harrisburg to Pittsburgh. Installation costs for this section were estimated to be about $52,000 per route-mile.[5] Since the survey simply publicized what the railroad already knew, it had little or no influence on the Pennsylvania's electrification strategy. Its impact upon other roads was nil as well, for the commission's report gave no hint that the federal government stood ready to help defray the costs incurred by any line that might wish to convert to electric traction. In the midst of the nation's most acute depression, with most roads marshaling all their economic resources just to ensure survival, expensive improvement programs supported solely by private capital were out of the question.

The Pennsylvania Railroad was in many ways the exception, of course. In 1936, the Clement administration commissioned Gibbs and Hill to study the feasibility of converting the Middle and Pittsburgh divisions to electric operation. In their report, completed in 1938, the consultants envisioned vast improvements for the line. They recommended boring several new tunnels in the Gallitzin area to reduce the steepness of the grades, for example.[6] Gibbs and Hill advised the PRR to step up the contact voltage to 50,000 volts and to outfit 20 new GG1's with oversized transformers, thus enabling the locomotives to work east or west of Harrisburg. Such a proposition initially would be expensive, claimed George Gibbs, but it would later result in significant economies of current distribution. The general increase in transmission efficiency that accompanied the use of higher voltages would decrease the number of substations needed and would lessen the amount of copper required in the catenary.[7]

The Pennsylvania's engineers opposed this idea, not only because it necessitated a higher outlay for motive power, but also because they questioned the practicality of installing the oversize transformers in the GG1's. General Electric and Westinghouse experts could give no assurance that the transformers could be built within the dimensions of that locomotive. As an alternative, the railroad outshopped a single experimental 5,000-horsepower locomotive for both freight and passenger service, Class DD2. This 2–B+B–2 unit incorporated the articulated running gear and welded, streamlined carbody of the GG1 and the high axle loads and horsepower common to the 01 and P5a classes. Internally, it contained nothing that was innovative. When subsequent testing revealed a number of deficiencies, however, the PRR's engineers decided that designing an entirely new locomotive type would be cheaper than trying to eliminate the DD2's flaws.[8] It was to be the last class of electric motive power that owed its design primarily to the railroad's own engineering staff.

Even had the DD2 been successful, the need for it was not urgent. The consultants issued their findings just as the national economy sank back into the doldrums after showing a brief upswing in 1937. The Pennsylvania decided that it could not afford the elaborate improvements recommended in the study and temporarily shelved the

A rare portrait of the PRR's single DD2, No. 5800, at the Wilmington shops, 1939. (William E. Grant)

whole project. In 1939, the PRR and General Electric jointly scrutinized the venture, but again the railroad concluded that the time for another extension of catenary had not yet arrived.[9]

Two years later, the Pennsylvania requested Gibbs and Hill to examine the matter a second time. Business was good in 1941, and in the not unlikely event of the United States becoming involved in another world conflict, electrification would be in the national interest as well. This time the PRR wanted the consultants to confine their planning to conversion of the line from steam to electric traction. The railroad had no money for new tunnels, relocations of rights-of-way, or other related improvements. In their new study, Gibbs and Hill calculated the cost of a "simple" electrification—exclusive of any upgrading of the line—to be $120 million. This included the erection of catenary over the main line between Enola and Conway yard (west of Pittsburgh) and along the Conemaugh division, a double-track freight route which diverted traffic to the north of Pittsburgh's crowded downtown area. If the PRR was willing to forgo electrifying the Conemaugh division and to substitute wooden for steel poles in the catenary structure, it would save $25 million, said the consultants. In either case, the price included the cost of new motive power and the utilization of 11,000-volt current. Using 1939 as a base year for normal peacetime traffic, Gibbs and Hill predicted that electrification would save the railroad $3.5 million annually in operating costs, not to mention more millions in maintenance costs. The faster service which the Pennsylvania could offer to shippers and travelers would bring more business to the railroad, too.[10]

The war intervened before the PRR had a chance to pass judgment on the consultant's findings. As in 1917, the world conflict and its resultant shortages and high prices put a stop to all major capital improvement programs.

The Clement administration still hoped that once the war ended, it would be able to proceed with the long-planned extension of electrification. Late in 1943, through a subsidiary, the Manor Real Estate and Trust Company, the railroad purchased 2,650 acres of prime coal land in Cambria county adjacent to the main line of the Pittsburgh division. Company officials could neither confirm nor deny the speculation which described the transaction as the first step toward electrification. Rumors circulated to the effect that the Pennsylvania was going to build a mammoth power plant at Wilmore to generate electricity for the trains and nearby industries. In 1944, both General Electric and Westinghouse were drawing up plans for two new types of electric locomotives for service between Harrisburg and Pittsburgh. One design, almost completed, was for a 7,500-horsepower 2–C+C–2, a kind of super GG1 that would be equally at home east or west of Harrisburg in fast freight or passenger service. The other design, still in the preliminary stages, called for a 10,000-horsepower giant whose duties were to be restricted to freight service west of Enola. The railroad's decision in 1945 to have its consultants update construction costs from their 1942 study lent further credence to the PRR's eagerness to expand its electrified territory. Gibbs and Hill found no appreciable differences in either costs or savings from the earlier investigation.[11]

In the years following the war's end in 1945, however, a number of difficulties arose to prevent the Pennsylvania from carrying out its plans. In 1946, the railroad suffered the first net deficit in its 100-year history, ending the year a dismal $8 million in the red. Based upon the volume of traffic, 1946 should have been a good year for the PRR. Gross operating revenue for the year was $758 million. Ton-miles totaled nearly 57 billion and passenger-miles almost 10 billion. In 1940—the last full year of peace—gross operating revenues amounted to only $435 million. Ton-miles totaled only 40 billion and passenger-miles 3.4 billion. Yet in 1940 the railroad earned a net profit of $46.2 million.[12] On the other hand, operating costs had risen dramatically between 1940 and 1946. With the cessation of hostilities had come a period of extreme inflation. Demands for peacetime goods and services, stifled for five years, now outstripped the economy's ability to satisfy them. Labor and material costs skyrocketed. The Pennsylvania had 10,000 fewer people on its payroll in 1946 than in 1945, yet it paid $480 million in wages in

1946 as compared to $449 million in 1945. The railroad claimed that the government had not permitted rate increases proportional to the cost of doing business, which further aggravated the line's financial plight.[13]

The PRR also blamed the excess profits tax which Congress had enacted in 1943, as a contributory factor in its poor showing for 1946. In 1944, President Clement wrote that the tax prevented the railroad from putting away monetary reserves "for future expenditures which are being made inevitable by wartime conditions and the pressure of wartime traffic. Thus deferred maintenance is mounting at an ever increasing rate. Obsolescence is accumulating."[14] Deferred maintenance, obsolescent equipment, and inflation resulted in increasingly inefficient operation. The railroad posted an operating ratio of 70.9% in 1940. Six years later, it had risen to 90.7%.[15]

Ironically, some business analysts contended that part of the responsibility for the Pennsylvania's predicament rested with electrification. Most of the road's capital improvement funds during the 1930s had gone for electrification and related projects. The PRR halted development of more efficient steam locomotive types. The railroad believed that the decade's light traffic levels and limited earnings did not warrant spending money on new steamers, particularly when electrification all the way to Pittsburgh appeared to be a certainty. Consequently, the ballooning traffic of World War II caught the Pennsylvania without a modern steam design. The railroad had to borrow blueprints for a 2–10–4 from the Chesapeake and Ohio in order to rush a class of superpower steamers into production. The bulk of the road's steam fleet nonetheless dated from the 1920s and even earlier and was very costly to operate under rigorous wartime conditions.[16]

The financial picture continued to be gloomy in 1947, although the road did manage to show a net imcome of $7.3 million for the year. Better still was 1948, when the railroad reported a net income of $34.5 million. But once inflation had been taken into account, even this sum appeared anemic in comparison to the rich harvests of the 1920s and barely equaled the depression incomes of the 1930s. Conversely, the cost of materials and supplies had risen an average of 105% between 1939 and 1948, and wages had gone up an average of 70%.[17]

In spite of the financial squeeze, the Pennsylvania did scrape together enough money in 1947–48 to dieselize its long-distance passenger trains operating between Harrisburg and the west—a sure sign that the railroad's enthusiasm for extending electrification had cooled. Unlike many other lines, the PRR chose not to experiment with diesel locomotives in the prewar years. What few diesels were on the roster were small switcher types. As late as 1945, the Pennsylvania clung to steam and had outshopped several new classes of engines in a tardy effort to increase the efficiency of steam locomotion. These attempts to build coal-burning locomotives that could match the performance of diesels were largely in vain. They only intensified the inefficiency which plagued the railroad's steam-driven motive power. Many observers blamed the PRR's faith in electric traction (along with its loyalty to the coal industry) for delaying the introduction of the diesel locomotive to the system. Had the road launched a dieselization program around 1940 and resumed it immediately after the war's end, it would not have been saddled with so many aging steam engines. Operating costs would therefore have been much lower. It eschewed early dieselization, however, and by the end of 1948 still had over 4,100 steam locomotives on its roster.[18]

The diesel locomotive did as much as the Pennsylvania's economic distress to obstruct the further growth of electrification. Each locomotive contained one or more diesel engines which, when coupled to d.c. generators, furnished electricity for traction motors. In this sense, the diesel locomotive was nothing more than an electric locomotive that carried its own power plant—hence the more accurate (but less widely used) term "diesel-electric." The diesel-electric offered many of the same advantages as a "pure" electric: continuous torque, high tractive effort, rapid acceleration, freedom from dependence on coal wharves and ashpits. In addition, the railroad could obtain the benefits of dieselization as soon as it put the new locomotives into service. There was no need to wait for two or three years while catenary was erected. A dieselization program could be implemented gradually, according to the road's financial resources; electrification could be accomplished only on an all-or-nothing basis involving a huge cash outlay. And a typical electric locomotive's cost per horsepower was substantially higher than that of a diesel locomotive. This cost differential stemmed largely from the economies of mass production that were available to the diesel manufacturers. Railroads were buying thousands of diesel locomotives in the late 1940s in an effort to rid themselves as soon as possible of the now-obsolete steam engines. In contrast, only the Great Northern, the Virginian, and the Milwaukee Road were purchasing electrics—and their combined orders totaled less than 30 units.[19]

By 1947, the Pennsylvania had to choose between enlarging its electrified territory or beginning massive dieselization. The railroad could either string wires west to Pittsburgh or dieselize the nonelectrified portions of its system. It could not afford to do both. Even if catenary did reach Pittsburgh, however, dieselization of the rest of the system was imperative. The PRR therefore decided that in the long run use of diesels would be the less costly and more economically rewarding of the two alternatives. By the end of 1947, the railroad had in service or on order 358 diesel locomotives. Long-distance passenger trains operating west of Harrisburg and helper service between Gallitzin and Conemaugh were among the initial targets for the new machines.[20]

By December 31, 1948, the PRR's directors had authorized acquisition of a total of 590 diesels, 352 of which had already been delivered. The performance of these new locomotives pleased the railroad. "The ton-miles of service per locomotive-hour produced by these engines so far surpasses the performance of any previous type of motive power," stated the company's annual report for 1948.[21]

The PRRs dieselization program soon surpassed electrification as the road's most expensive single investment. The bill for almost 600 diesels, plus maintenance and repair facilities for them, stood at $170 million in 1949.[22] Total cost of the project when officially completed in 1957 was $400 million. Adjusted for inflation since 1935, this sum was roughly equivalent to the $200 million or so that the PRR had spent on electrification and related improvements since 1928.[23]

The Pennsylvania's decision not to electrify its Middle and Pittsburgh divisions did not eliminate the railroad's need for new electric motive power. The PRR was less than satisfied with the P5a as a freight hauler. The railroad had regeared the P5a's to a maximum speed of 70 miles per hour, but this alteration did not entirely cure their adhesion difficulties.[24] Moreover, a single P5a's continuous tractive effort at low speeds was inadequate to handle many freights. The PRR was forced to double-head the engines on heavier trains, a wasteful practice that electrification was supposed to eliminate. This is not to say that the P5a's were failures as freighters. They performed admirably in merchandise service, where tonnage levels were not so burdensome and where speed was essential. Yet the Pennsylvania had enough surplus GG1's to handle these fast runs. The operating department had little choice, therefore, but to relegate the P5a's to hauling heavy tonnage such as coal trains, even though the locomotives were not ideally suited for that type of work.[25]

The PRR originally figured that a sufficient number of new electric engines would be constructed in connection with the electrification to Pittsburgh to supplement the P5a fleet as well as to handle assignments in the newly electrified territory. With plans for the extension now dropped, the railroad still faced the expensive prospect of finding a successor to the P5a. At first, the road's motive power planners favored completing the development of the 7,500-horsepower locomotive, the design for which was fairly well advanced. The Pennsylvania's studies of 1946 indicated that 21 of these giants would be needed to relieve the problem of double-heading the P5a's, mainly on eastbound runs from Enola where trains encountered the

P5a's No. 4718 and 4725 approaching River Interlocking, Baltimore, with a coal train, 1961. (Russell L. Wilcox)

22-mile, 0.3% Smithville grade between Columbia and Parkesburg.[26] In the spring of 1947, railroad engineers drew up a list of general specifications which GE and Westinghouse were to follow in their designs. Besides clearance and weight restrictions, each locomotive had to be capable of pulling a 10,000-ton, 125-car train on level track at 50 miles per hour, or up a 0.5% incline without exceeding its continuously rated horsepower (in other words, without overloading its traction motors). Maximum continuous tractive effort had to occur at or below 22 miles per hour.[27] The PRR already owned an incomparable racehorse in the GG1. It now demanded a locomotive that could efficiently pull slower and heavier trains, particularly coal drags.

The railroad never placed an order for any of the 7,500-horse-power engines, however. Probably the company's financial reverses partially explained this inaction. And certainly the road wished to see what the new diesels could do before it made a sizable investment in new electric power. There was little doubt that the Pennsylvania needed more modern motive power east of Enola. The railroad had expected traffic to decline rapidly to prewar levels after 1945. In fact, business fell off only slightly, to both the delight and consternation of management. In December 1948, Vice-President of Operations James M. Symes advised President Clement that the PRR had to dispatch as many as 40 trains per day in the electrified region behind steam power owing to the shortage of electric engines. Complaining that "the Pennsylvania Railroad has never had a satisfactory electric locomotive design for freight service," Symes urged Clement to have the railroad acquire several experimental units, test them thoroughly, and then purchase at least 20 of the most suitable types.[28] The PRR's president was already eager to begin work on developing new electric locomotives for his road. Earlier in the year, he had written to Charles E. Wilson, president of General Electric, stating that "there has been no real progress in single phase electrific propulsion since the GG1 was built" and suggesting that railroad and GE engineers get together to discuss "the possibility of making some progress."[29]

Clement's willingness to consider purchasing new electrics prompted General Electric and Westinghouse to conduct their own assessment of the PRR's motive power situation. L. W. Ballard, GE's engineer in charge at Philadelphia, informed the railroad that his company's studies showed that a 90-car coal train headed east from Enola would require diesel locomotives totaling 6,000 horsepower and carrying a price tag of $617,000. One 5,000-horsepower electric locomotive, costing only $452,000, could haul the same train. The studies also revealed that diesels would operate at much slower speeds, especially when climbing grades. A slow-moving coal train might very possibly delay faster trains which followed it, causing traffic to back up at the terminals. The surveys done by Westinghouse reached much the same conclusions.[30]

Although the rated horsepower of a diesel locomotive might equal that of an electric, its horsepower as delivered at the rail was significantly less. The diesel locomotive's drawbar capacity was determined by the prime mover's (that is, the diesel engine's) ability to drive the generator, thus creating electricity. Since the prime mover could not be overloaded, a definite limit existed as to how much current the generators could produce. In addition, some horsepower (approximately 20%) was lost in changing mechanical energy to electrical energy and back to mechanical—hence the need for a 6,000-horsepower diesel as opposed to a 5,000-horsepower electric. The prime mover for the electric locomotive was located at the central power station and could furnish a virtually unlimited supply of current to the traction motors.[31]

By 1949, the Pennsylvania had finally assured itself of the continuing superiority of electric traction. Railroad engineers met with representatives from General Electric, Westinghouse, and Baldwin in August of that year to discuss certain revised clearance, weight, and safety standards which any new locomotives would have to meet. Other than these basic guidelines, the PRR gave the builders a free hand in designing the new machines. Orders were placed with each manufacturer for four experimental units.[32]

The studies of the railroad's electric locomotive operations done by GE and Westinghouse the previous year evidenced that both manufacturers had scrapped plans for a 7,500-horsepower unit. Instead the engineers at Erie and East Pittsburgh had decided that merely producing a larger version of the GG1 would not be in the manufacturers' best interests, nor those of the Pennsylvania.

General Electric and Westinghouse realized that one of the chief barriers to the growth of railroad electrification in the United States was the high cost of motive power. If they could construct units of relatively low cost for the Pennsylvania, perhaps other roads would more seriously consider converting to electric traction or at the very least replace obsolete electric locomotives with new ones. The PRR's test units, therefore, had to be engineered not only to satisfy that railroad's specific needs but also to fulfill the general service requirements of other lines. This line of thinking represented a notable departure from the builders' past (and expensive) practice of custom-tailoring locomotives to meet the specific demands of individual railroads. The best way to cut costs and to ensure their products' versatility, GE and Westinghouse engineers reasoned, was to take advantage of the technology and the popularity of the diesel locomotives. Baldwin, Alco, Fairbanks-Morse, and especially General Motors' Electro-Motive Division (EMD), by exploiting the advantages inherent in mass production, were turning out diesels that were relatively low in cost. If the Pennsylvania's new electric locomotives could incorporate some basic designs from diesel technology, their purchase price and maintenance costs could be lowered substantially.[33] The techniques of diesel production were not unfamiliar to either of the electrical manufacturers anyway, since GE built electrical components for Alco, while Westinghouse provided similar gear for Baldwin and Fairbanks-Morse.

In June 1951, the PRR took delivery on two two-unit electric freight locomotives from General Electric. The railroad designated the new engines "Class E2b" and gave them road numbers 4939–4942. Except for the pair of pantographs mounted atop each unit, the new machines bore a striking resemblance to diesel road locomotives. At one end of each unit was a cab fronted by a rounded, bulldog nose. Two units coupled back to back constituted one double-cabbed locomotive. Each unit rode on two four-wheel trucks. Again in imitation of diesel practice, all four axles received power, giving the units a B–B wheel arrangement. General Electric engineers contended that the cost per useful pound of weight for the P5a and GG1 was too high, since nonpowered axles carried some of the weight. In the E2b's, all weight rested on drivers and therefore added substantially to the locomotives' tractive effort. Each locomotive weighed 245 tons and developed 5,000 continuous horsepower.[34]

Initially GE considered building units having the same kind of

Two-unit E2b Nos. 4941-42 on the test track at GE's Erie Works in 1951. (General Electric Co.)

axle-hung, d.c. traction motors that all diesels used. The PRR would then be able to service the new electrics at the same facilities where it serviced diesel locomotives, thus lowering maintenance expenses even further. For applications where a high continuous tractive effort was desired, especially in the lower speed ranges, d.c. motors were more efficient than their a.c. counterparts. Direct current motors also required less maintenance, primarily because their more uniform torque lessened the strain on bearings and other components. However, motor-generator sets offered the only reliable means by which the alternating current from the contact wire could be converted to direct current for use by the traction motors. Motor-generator locomotives had done well enough—GE only recently had outshopped m-g units for the Great Northern and the Virginian—but they were exceedingly heavy and very expensive. The use of m-g sets did not mesh with General Electric's attempt to attract new customers to the electric locomotive market. The PRR did not care for motor-generator locomotives, either. They cost far too much; and in order to satisfy weight restrictions, they would have to ride on a good many axles, a characteristic which the railroad's motive power experts traditionaly disliked.[35]

So GE remained true to alternating current. It equipped each E2b unit with four Type GEA–632 traction motors, similar in design to

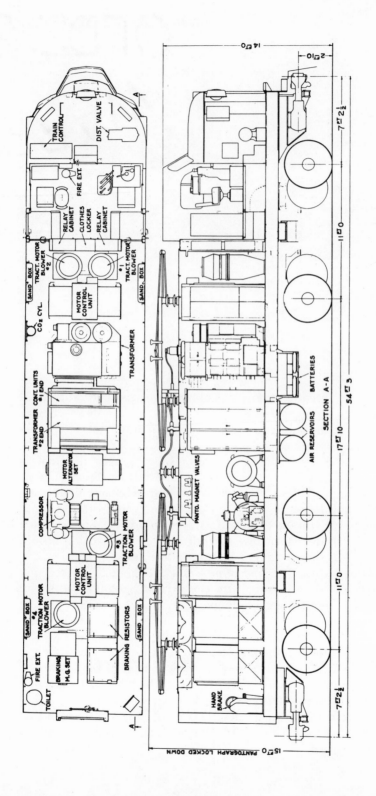

Interior layout of an E2b. (General Electric Co.)

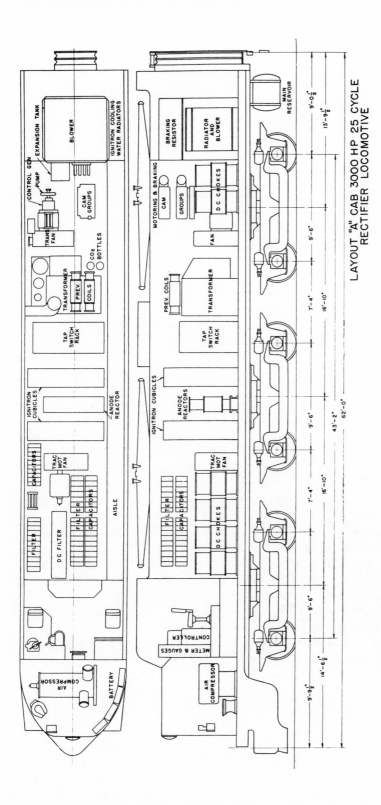

LAYOUT "A" CAB 3000 HP 25 CYCLE
RECTIFIER LOCOMOTIVE

Interior layout of an E3b. (Westinghouse Electric Corp.)

the Type 625–A motors found in the P5a.[36] To be sure, a number of advantages attached themselves to the use of straight a.c. locomotives, as GE's project engineers explained.

> The straight a.c. type locomotive . . . has been the basic motive power on the Pennsylvania's electrified lines since 1931 and has given commendable service. Railroad schedules, operating practices, and shop facilities have been built around it. Twenty years' experience with this plant has demonstrated that simplicity of mechanical construction and electric circuits is the key to low maintenance and reliable operation.[37]

Several months after the E2b's arrived, they were joined by two two-unit sets of locomotives from Westinghouse (with carbodies and mechanical gear supplied by Baldwin). In external appearance, the Westinghouse machines were nearly identical to their GE competitors. Internally, there existed extraordinary differences. The primary distinction lay in the utilization by the Westinghouse models of ignitron rectifiers to change alternating current to direct current. Westinghouse had advanced one bold step further than General Electric in the battle to evoke greater interest in railroad electrification. A locomotive drawing alternating current from overhead but using d.c. traction motors would not only possess certain technical advantages over a straight a.c. machine; it would be in a much better position to capitalize on the economies of diesel production, since many of the components would be identical.

The adoption of rectifiers for railway service was not entirely novel. Both GE and Westinghouse experimented with them for a brief period prior to World War I. In 1908, in the first recorded use of a rectifier for motive power purposes, General Electric outfitted with a mercury arc rectifier a car belonging to the Paul Smith's Electric Railway of Paul Smith's, New York. Five years later, Westinghouse installed a similar device in a New Haven m.u. car. None of these early tests met with much success, so both firms began concentrating their energies on developing motor generators instead.[38]

Despite the initial setbacks, at least one man at Westinghouse did not abandon faith in rectifiers. Lloyd J. Hibbard, just a cub engineer when he worked on the New Haven experiment, never wavered in his conviction that rectifiers could be perfected for railroad use. By the time World War II ended, Hibbard had begun his campaign to have his employer conduct another series of tests with the rectifier. The rectifier itself had undergone considerable improvement over the years. The ignitron had succeeded the mercury arc in the 1930s. It was further refined during the war, when the steel and manganese industries depended upon ingitron rectifiers to furnish large quantities of direct current.[39]

Hibbard's persistence paid off. In 1949, Westinghouse requested and received permission from the PRR to place an ignitron rectifier in one of the road's standard MP54 m.u. cars. On July 15, 1949, MP54 No. 4561, a seemingly ordinary passenger-baggage combine, emerged from the railroad's Wilmington shops outfitted with a set of ignitron tubes.[40]

The electrical circuits within each tube were designed to permit current to flow through in one direction only. Thus, when the alternating current collected by the pantograph was shot into a tube, all of the "back and forth" (alternating) impulses of the current were rectified, or made to flow in the same direction. By connecting two tubes, both the positive and the negative aspects of the cycle could be rectified and direct current produced. Additional equipment then "smoothed out" the direct current before it was fed to the traction motors. The ignitron itself was a comparatively simple device and occupied only a small space. (Each tube measured slightly less than 4 feet in ~~diameter~~.)[41]

For the next several months, Westinghouse, PRR, and Bell Telephone engineers closely monitored the performance of No. 4561. The Bell people had been called in because in the past, rectifiers had shown a disturbing tendency to interfere with the functioning of nearby telephone circuitry. Fortunately, only minor problems of this nature arose and were quickly remedied by the technicians. In fact, no serious difficulties of any sort manifested themselves during the test period. Some skeptics had predicted that the mercury pool in each tube would slosh so violently once the car was in motion that a continuous electrical circuit would never be achieved. Others wondered just how long a vacuum could be preserved in the tubes under normal operating conditions. These fears proved to be groundless. The trials went so flawlessly that after No. 4561 had run over 2,500 miles on direct current, the Pennsylvania once again placed it in regular revenue service.[42]

The railroad expressed complete satisfaction with the results of the equipment. Westinghouse, too, was delighted, for the successful trials of the m.u. cars represented conclusive evidence of the feasibility of applying ignitron rectifiers to locomotives.[43] The builder forecast that its new ignitron rectifier locomotives would offer all the qualities that the Pennsylvania wanted in a new freight hauler. Westinghouse engineers pointed out, for example, that a rectifier-equipped locomotive would develop an average of 50% more tractive effort at low speeds (under 17 miles per hour) than would a straight a.c. type of comparable horsepower. At 40 miles per hour, the tractive effort of the two engines would be about the same. Above that speed, the straight a.c. locomotive would enjoy a slight but not significant advantage.[44] Given the large volume of coal and other bulk commodities which it carried, the Pennsylvania must have considered the rectifier units a most inviting proposition.

Numerous other advantages accompanied the use of rectifier locomotives. They had far fewer moving parts than either straight a.c. or diesels. The rectifier apparatus was relatively lightweight (weighing about one-fourth as much as a motor-generator set) and required less space than a.c. equipment. And of course the railroad could not afford to overlook the superiority of the d.c. traction motor, which could be used with rectifiers. Of great significance for the future was the ignitron's ability to operate at either 25 or 60 cycles. Utilities generated most of their power at 60 cycles. The PRR could realize substantial savings by switching to 60 cycles in the event it upgraded or expanded its electrified lines, since it would not have to rely on expensive frequency-changing equipment. While older motive power would be virtually useless if 60-cycle current were adopted, the ignitron-rectifier locomotives would require only a slight, inexpensive modification.[45]

In the fall of 1951, Westinghouse and Baldwin delivered their first two-unit locomotive to the Pennsylvania. Classed E2c (Nos. 4997–4998) by the railroad, each unit had a C–C wheel arrangement. The locomotive was heavier (by 117 tons) and more powerful (by 1,000 continuous horsepower) than the E2b, and also cost more ($680,000 for an E2c, $500,000 for an E2b). Each unit housed 12 ignitron tubes, or 2 for each traction motor. In February 1952, the PRR received a second, nearly identical set. The major difference was the B–B–B trucks on which each unit rode, which led the railroad to designate this locomotive E3b (Nos. 4996–4997). Both the E2c and the E3b employed Type 370 DZ traction motors, the same kind featured on most Baldwin and Fairbanks-Morse diesel locomotives.[46]

At first, the railroad was pleased with the performance of all its new experimentals. PRR Senior Electrical Engineer S. V. Smith reported that after several months of trials in revenue service, "both types of locomotives had indicated that they are capable of hauling freight more efficiently than the older electric locomotives or than a diesel-electric locomotive of comparable size."[47]

A favorite assignment for the new engines during their early days was the 132-mile run between Enola and Morrisville. A single P5a carried a rating of some 4,500 tons on this route. A GG1 was rated at 4,280 tons. Experience quickly showed that the two-unit E2c could easily pull a train in excess of 12,000 tons. On February 19, 1952, for instance, the E2c pulled a 160-car coal train weighing 13,348 tons from Enola to Morrisville and set a new tonnage record for the line.[48] A few months later, sister E3b was accomplishing similar feats. In other words, the railroad needed three P5a's or three GG1's to do

E2c No. 4997 at South Philadelphia, 1954. (Harold K. Vollrath Collection)

the work of one two-unit rectifier. Road tests also demonstrated that a P5a would haul its maximum-rated load from Enola to Morrisville in 4 hours, 50 minutes (under optimum conditions). Rectifiers bettered this time by a half hour. Neither locomotive could match the GG1's dash of 3 hours, 28 minutes over the line; but an E2c or E3b could pull a train three times the weight of a P5a's or GG1's maximum allowable tonnage. This meant that the rectifiers could produce about three times the gross ton-miles per train-hour than either of their older competitors. While the E2b's proved to be a little faster than the rectifiers, their 7,800-ton rating (for two units) on the Enola-Morrisville run was less than that for a pair of P5a's or GG1's.[49]

The Pennsylvania declared in April 1952 that its three experimental classes were performing "creditably" and were encountering only "insignificant" maintenance problems. Even excessive lateral thrust, that persistent plague of earlier PRR electrics, posed little threat.[50] The locomotives were not used in high-speed service, where side blows became a problem. Experience gained in perfecting the suspension systems of diesel locomotives helped alleviate the problem as well. Soon the test models joined the 152 other electrics in the railroad's freight pool. Although all three classes saw duty in almost all kinds of freight operations, the PRR used them whenever possible in drag freight service, since that was the work for which they had been designed.[51]

E3b No. 4996 leading an E2c and two diesel units on a southbound freight near Wilmington, 1956. (Fred W. Schneider III)

The experimentals' successes lasted but a short time, however. None of the locomotives could cope with the rigors of heavy-duty railroading. The number of road failures increased sharply. Breakdown after breakdown transformed the engines into mechanical and electrical nightmares for the PRR's maintenance forces. The railroad had withdrawn all the units from service by mid-1954 and stored them at the Wilmington shops pending major overhauls.[52]

When actually operating, the locomotives compiled a fine record of achievement; but keeping them in operable condition was by no means an easy or inexpensive task, as S. V. Smith recounted some years later.

> The units from Westinghouse suffered from the fact that they were the first units to incorporate rectifier apparatus in revenue service in the world and were built with a poor layout design by Baldwin and Westinghouse. Among other things the poor layout was due to inexperience with rectifier apparatus and also to an attempt to make the locomotives readily convertible to a straight a.c. type if the rectifier plan did not prove workable. This poor layout resulted in a locomotive that is difficult to maintain . . . and a constant drain on maintenance budget. With the exception of the initial period when these units were closely followed by engineering representatives of the builders, all units have given poor performances.[53]

A General Electric engineer who later had occasion to examine the E2c's internal apparatus remarked that the layout resembled "a country hardware store" more than the interior of a locomotive[54]

The E2b's performed only slightly better than their Westinghouse rivals. They were particularly prone to traction motor failures. While General Electric had based the Type 632 motor upon proven designs, it was much larger than earlier motors. Packing such an unprecedented amount of horsepower into a single motor caused excessive heat buildup and premature component wear.[55] L. W. Ballard of GE worried that "our series of difficulties with the GEA 632 motors has . . . seriously undermined the customer's confidence in the possibility of building a satisfactory single-phase a.c. motor of that rating."[56] These fears were well founded. After the E2b's had been in service several years, the PRR's chief mechanical officer, L. E. Gingerich, stated flatly that "we would not consider purchasing similar units."[57]

In fairness to Westinghouse and General Electric, their locomotives' problems did not result exclusively from design defects. The railroad was also at fault. In 1954, the Pennsylvania asked GE to study the road's electric locomotive maintenance program and suggest ways to reduce expenses. Maintenance costs for electric locomotives had risen from an average of 12.2 cents per mile in 1940 to 62.8 cents per mile in 1954.[58] In its subsequent report, GE blamed much of the increased cost on inflation, but also criticized the quality of inspection and repair which the PRR was giving to its electric fleet. The railroad's shops were maintaining equipment having a counterpart in steam construction (brake shoes, wheel tires, journals, springs) to a much higher standard than the electrical equipment, but even the more familiar components were not receiving the same degree of care that had prevailed in past years.[59] The railroad's poor financial showing since the war was beginning to take its toll, as reduced and deferred maintenance became ever more widespread.

The Pennsylvania did not order duplicates of any of these E-class experimentals. The rectifier units particularly had shown great potential, but the railroad was dismayed at their high cost of operation. The PRR had obtained valuable experience with rectifiers which undoubtedly would contribute to lower maintenance bills for production models. Nevertheless, the railroad was not sure that these costs could be kept within acceptable limits.[60] Further inhibiting the road's desire to place an order for more electrics was the company's chronic financial weakness. The combination of unsatisfactory revenues and the need to use what little cash that was on hand for urgent projects all but closed the door to any serious consideration of enlarging the electric fleet in the near future.

The Pennsylvania was not the only railroad to display a keen interest in rectifier-type motive power. The New Haven, impressed by the early triumphs of the E2c and E3b, placed an order late in 1952 for 100 ignitron-equipped, stainless steel multiple-unit cars. Westinghouse was to furnish the electrical gear, and Pullman Standard would supply the carbodies and mechanical parts. The New Haven was also considering the acquisition of ignitron-rectifier locomotives, as was the Virginian Railway, which was using split-phase box cab units 30 years old and in dire need of replacement. In 1954 the New Haven ordered 10 ignitron-equipped, 4,000-horsepower passenger engines (Class EP-5). The following year the Virginian announced that it was buying 12 3,300-horsepower electric freight locomotives employing ignitron rectifiers (Class EL-C).[61]

Westinghouse did not receive the locomotive orders, however. Disappointed with the lack of interest in railroad electrification since the war, the builder had withdrawn from the heavy electric traction field in 1954.[62] (The Pennsylvania's failure to order duplicates of the E2c and E3b may well have played a crucial role in Westinghouse's decision to leave a market that had been shrinking for 20 years.) Thus General Electric became the nation's sole supplier of heavy-duty electric locomotives.

General Electric's engineers had not closed their eyes to the possibilities of developing rectifying apparatus for railway use. They simply believed that rectifiers required further improvement before the devices could profitably be placed in locomotives. When the PRR's m.u. car had been fitted with ignitron tubes, L. W. Ballard warned the head of his firm's Transportation Engineering division that "it is very important that we keep an eye on the Westinghouse development." D. R. MacLeod, a guiding force behind GE's E2b's, admitted the ignitron-rectifier locomotive as built by Westinghouse to be "a very worth-while contribution to the art of electrification," and said that GE's engineers would "watch its evolution on the PRR with sympathetic interest."[63] In January 1953, with its own rectifier research and development well advanced (and possibly forewarned of Westinghouse's decision to quit the business), General Electric established a committee to begin planning specific applications of rectifiers for use in locomotives. Henceforth the company would market the rectifier as a superior alternative to either m-g sets or straight alternating current.[64]

Evidence of the success of this approach could be seen in the New Haven and the Virginian transactions. On the other hand, General Electric made no immediate sales to the PRR. In 1955, that railroad entered a four-year period during which it evaluated its entire electric traction operation. The purchase of new motive power (diesel or electric) for the lines east of Enola was accordingly postponed.

The mid-1950s showed an economic picture little brighter for the Pennsylvania than that of the late 1940s. Net railway operating income for 1954 totaled only $48 million—a smaller amount than any recorded during the depression years. The PRR amassed fewer ton-miles (42.3 billion) for 1954 than for any year since 1940. Passenger-miles had fallen to 3.8 billion, again the lowest figure recorded since 1940. Business picked up a bit over the next two years; but in 1957 net income once more dipped sharply, to a mere $19 million.[65]

The causes of the financial difficulties experienced by the Pennsylvania after World War II remain a matter of some controversy today. There is no question that one of the most fundamental problems facing the railroad was the rising cost of doing business. In the 15 years between 1942 and 1957, for example, hourly wages rose 152%, the cost of an average boxcar 102%, and the price of steel rails 166%. Coupled with what the railroad claimed to be inadequate rate increases and unfair government subsidization of nonrail competition, inflation crippled the company's earning power.[66] Yet these ills afflicted all railroads. Perhaps of greater consequence in explaining the PRR's own predicament were the high terminal costs that characterized the eastern end of the system, the deficit-riddled passenger service, and most importantly the general decline of business—especially the coal industry—in the northeastern United States. The road's management capabilities (or lack of them) have also come in for a share of the blame. Martin W. Clement stepped down from the presidency in 1949 and was succeeded by Walter S. Franklin. Upon Franklin's retirement in 1954, the board of directors elected James Symes as president. All three men had spent practically their entire careers with the PRR, and all have been criticized as typifying the kind of provincial mediocrities that corporate inbreeding so often fosters. How accurate that criticism may be is not at issue here. The point is that the magnitude of the external factors weighing against the prosperity of the Pennsylvania was such that the road would have suffered serious hardship after 1945 regardless of how expert its leadership was. Therefore, any proposed enlargement or modernization of electrified service was bound to be severely constrained, even under the most able group of managers.

"We are deteriorating badly," a grim James Symes told a Congressional subcommittee in 1958.[67] The railroad in 1955 had completed a billion-dollar improvement program that included dieselization, modernizing and enlarging yards, building a new shop complex (the world's largest) at Hollidaysburg, and erecting a huge iron ore unloading pier at South Philadelphia. Since his company was "not a very attractive investment," Symes stated, much of the money to finance these improvements had come from selling nonrail properties and deferring maintenance. While such a course of action provided ready cash, it was bound to have an adverse effect upon the railroad in the long run. The PRR was able to replace only 6,200 tons of rail annually in recent years, testified Symes. Under normal economic circumstances, it would replace 50,000 tons. The Pennsylvania's president reported that his road was making every effort to slash operating costs, boost productivity, and heighten the overall efficiency of his physical plant. The billion dollars' worth of improvements would help some in this regard, but by themselves the improvements would not be sufficient to prevent an eventual financial collapse.[68]

As an additional improvement, the PRR had briefly considered expanding the electrified territory. In May 1955, President Symes announced that Gibbs and Hill, in cooperation with General Electric, was examining the possibility of extending catenary from Enola as far as Altoona. A 1947 study by the railroad had indicated that the Middle division could be electrified without requiring the purchase of a large number of new electric locomotives. If that still held true ten years later, electrification could release enough diesels for service elsewhere so that the Pennsylvania could delay the purchase of new units. Savings accruing from this measure, added to the known benefits of electric operation, might be substantial enough to warrant running the wires west to Altoona.[69]

Gibbs and Hill estimated that only a few electric engines (17) would be needed if the Middle division were electrified. The consultants nevertheless argued against the erection of any new catenary. "Extension of the electrification to Altoona or beyond can be justified only if the competitive situation demands higher speeds or if there is a substantial increase in the cost of diesel operation, including fuel oil."[70] In Gibbs and Hill's opinion, the numerous curves and grade crossings found on the Middle division would prevent the railroad from achieving anything but a marginal decrease in running times. This had not been the case 15 years earlier, when steam had ruled this portion of the main line; but the introduction of diesels had brought about a significant increase in freight train speeds. As for long-haul passenger traffic, it had fallen off to such a degree by the mid-1950s that the consulting firm no longer considered it to be an important element in the electrification question. And although fuel oil had doubled in price since 1945, diesel operation was only slightly more expensive, irrespective of the basis used for comparison, than electrified service would be. The Pennsylvania would save only about $1.5 million annually by electrifying. This was not a sum worth the $48 million initial investment, said Gibbs and Hill, even though some diesels (but not as many as the PRR anticipated) could be transferred to other parts of the system.

The consultants believed the railroad could more wisely spend its money on improving service in the region already electrified, where both passenger and freight traffic was still heavy. They urged the PRR to acquire new electric passenger locomotives (thereby releasing the GG1's for freight duty) and to replace its aging fleet of MP54's with stainless steel rectifier cars similar to those purchased by the New Haven.[71]

In 1957, General Electric secured permission from the Pennsylvania to conduct a motive power study of its own. The builder's subsequent investigations disclosed that the road's electrics were failing at a disconcerting rate. The GG1's were unavailable for service an average of 11.5% of the time, the P5a's 20%. The PRR had as many as 43 diesel units working under the wires on any given day, hauling trains that should have been electrically propelled. The addition of these diesels by no means solved the motive power shortage. During January 1957, a typical month, 324 trains in the electrified territory were delayed 3 hours or more for lack of locomotives of any kind. Like Gibbs and Hill, GE recommended that the railroad obtain new, more powerful locomotives for service east of Enola. In contrast to the suggestions of the road's consultants, General Electric argued that the Pennsylvania should award first priority to the modernization of its freight roster, since freight traffic earned considerable income for the road, while passenger service lost millions of dollars yearly. Even the acquisition of new locomotives could not eradicate the red ink that drenched passenger operations. However, GE's engineers made no specific recommendations as to how many or what kind of freight engines the Pennsylvania should buy, although they did note that the 3,300-horsepower rectifiers then being built for the Virginian would make excellent successors to the P5a's.[72]

The PRR, while acknowledging the need for more motive power east of Enola, was not convinced that the new locomotives should be electric. After receiving GE's report, Vice-President of Operations James P. Newell proposed to Vice-President of Transportation and Maintenance Allen J. Greenough that the railroad study the question of electrification's real value. Diesels had turned in a fine performance thus far. Rather than contemplate an extension of catenary, Newell suggested, perhaps the PRR should think about de-energizing it completely in favor of diesels.[73]

The railroad could not afford to delay this assessment of electric traction. Indeed, an upsurge in freight traffic caused in part by the road's newly developed iron ore business caused the motive power shortage to become so critical by 1957 that the PRR began shopping for second-hand electrics that could be used until a decision was reached one way or another on electrification. The Pennsylvania's engineers first examined the LC-2 engines that the Norfolk and

Western had put in storage after it de-electrified its Elkhorn grade in 1950. The condition of these units was judged to be so poor as to prohibit their economical return to service, so the search continued. In 1956, the Great Northern had de-energized its 74-mile electrification in the Cascade Mountains of Washington, an action that made surplus a number of well-maintained locomotives. For $222,000, the PRR purchased seven Class Y1's and a single Y1a. Built by GE in the late 1920s and rated at 3,300 continuous horsepower, these machines were giants even by the Pennsylvania's standards. They also constituted the first (and last) motor-generator power ever owned by the PRR. The Y1a was stored to supply parts for the Y1's, which after undergoing some modifications at Altoona emerged as Class FF2's. The FF2's were then placed in helper service westbound between Philadelphia and Paoli, both east and west from Thorndale, and eastbound from Columbia. None of these grades was especially steep, so most trains did not require helpers. Only the heaviest movements, mainly eastbound coal and westbound ore extras, demanded additional power, and then it was only to ensure that these trains maintained a high enough speed to prevent other traffic from being delayed. Helpers were also valuable in starting heavy trains that had been halted for any reason on these grades. In this capacity, the FF2's released over a dozen diesel units for other service.[74]

The outstanding job done by the FF2's did not alter the Pennsylvania's willingness to consider total dieselization, which was, indeed, becoming even more appealing. The power contract with Philadel-

FF2 No. 4 (formerly Great Northern Class Y1 No. 5014) at Thorndale, Pa., 1959. (Harold K. Vollrath Collection)

phia Electric expired in 1959. The utility had already stated that it wanted a rate increase in the new contract. Contracts with the other utilities were due to expire in a few years as well, and there was every reason to expect rate hikes across the board.[75]

Complicating matters even further was the possibility of a merger between the PRR and its one-time nemesis, the New York Central. Both roads were burdened with their share of difficulties. A union of the two companies, by ending duplication of facilities and services and by combining management strengths, would presumably yield a single financially healthy super railroad. At least that is what James Symes believed, and in 1957 he opened negotiations with the Central. A consolidation of the two lines could conceivably alter the Pennsylvania's traffic patterns to the point where a reduced volume of business would no longer justify electrified service on some of the present routes, let alone any new ones.

In 1958, the Pennsylvania contracted with General Electric and the two major diesel manufacturers—EMD and Alco—to conduct independent reviews of its electric operations. By mid-1959, all three builders had submitted their reports, and all three recommended without qualification that the PRR retain its electrification. When the railroad had electrified nearly 30 years earlier, it had reaped the benefit of the artificially low prices for materials and labor which the depression had spawned. Dieselization would have to be implemented in a period of inflation. And while diesels initially cost less on a per unit basis, past experience indicated that they would not enjoy the long service lives of their electric cousins without receiving intensive maintenance and frequent renewal of major components. This last point deserved special consideration by a railroad that since 1945 had been forced by rising costs and falling revenues to order severe retrenchments in its motive power maintenance program.[76]

Nor could diesels meet operational demands as well as the electrics. General Electric, for example, noted that the PRR derived 60% of its freight revenue from fast freight traffic, such as Truc-Train. Diesel locomotives, with their slower speeds, might well experience difficulty in meeting the road's most demanding schedules. Moreover, diesels could not be substituted for electrics on a one-for-one basis. The GE study estimated that 377 diesel units would be needed to replace the PRR's approximately 150 electric freight locomotives. Even with the cost of electricity rising slightly, General Electric calculated that dieselization could raise the Pennsylvania's operating costs east of Enola by as much as $21 million annually.[77]

Perhaps in a desperate attempt to find a compromise solution, the railroad early in 1959 investigated the feasibility of converting some of its diesels to straight electrics. Such a conversion was theoretically possible, now that a dependable rectifier had been developed. At first glance, the prospect was a most tempting one. The cost, the road's engineers calculated, would be about $72 per converted horsepower versus $110 per horsepower for a completely new electric locomotive. Further research eventually convinced the PRR that modifying the diesels would not be a sound alternative, after all. First, the road would have to buy new diesels to replace the ones it had changed to straight electrics. In addition, the solution would not be a long-term one, because the life expectancy of the hybrid units would not equal that of either new diesels or new electrics.[78]

On the basis of this study and those of the locomotive builders, the Pennsylvania decided not only to continue using electric traction, but to replace all of its P5a's, too. Vice-President Greenough directed that negotiations with General Electric be commenced promptly. The PRR's engineers had observed GE's 3,300-horsepower electrics in operation on the Virginian and had come away impressed with the machines' capabilities. The Pennsylvania preferred more powerful units, however, in view of the faster schedules which it maintained.[79]

In the fall of 1959, the railroad ordered 66 4,400-horsepower locomotives—to be designated Class E-44—from GE. This represented the largest order for electric engines by an American railroad since the PRR contracted for GG1's in the 1930s. Because the Pennsylvania could not raise enough cash to cover the $32 million price of the order, it could not purchase the E-44's outright. Rather, GE leased the locomotives to the railroad for a 15-year period, after which the PRR would become their owner.[80] The new units, with carbodies resembling those of a diesel road switcher, arrived at the rate of two per month, beginning in October 1960. On receipt of the first E-44 (No. 4400), A. J. Greenough, who had succeeded James Symes as president, said his company was "taking a long, confident look into the future as we acquire these locomotives. These units will serve for many years to increase the capacity of our electrified lines, improve service to shippers, and strengthen our reserve potential in the unhappy event of a national emergency."[81]

The first 36 E-44's employed the usual ignitron tubes to rectify the current. In July 1962, GE delivered the first E-44 (and the world's first locomotive) equipped with a silicon-diode rectifier. The silicon rectifier was a solid state device belonging to the transistor family. It soon proved to be much more efficient and reliable than the ignitron, mainly because it could do without the complex cooling system and firing circuitry of the ignitron. The silicon rectifiers had an initial failure rate only one-sixth that of the older tubes. Five additional E-44's were delivered having the silicon diode in place of the ignitron. Their performance was so satisfactory that the Pennsylvania

E-44 No. 4400, the first of 66 such units built by GE for the PRR. (General Electric Co.)

announced its intention to install the newer type of rectifier on all E-44's eventually. So efficient were the solid state devices that even the diesel manufacturers began to use a system of a.c. generators and silicon rectifiers to furnish direct current for diesel locomotive traction motors.[82]

The E-44's fulfilled the PRR's highest expectations. They could easily handle more tonnage and at higher speeds than the P5a's, while their greater tractive force allowed the railroad to discontinue the expensive practice of regularly stationing helper power along the east-west main line. The FF2's were withdrawn from service soon after the arrival of the E-44's, and by early 1963, when GE delivered the last of the new units, the Pennsylvania had already retired a sizable portion of its P5a fleet. By April 1965, the last of these sturdy 2–C–2's was stricken from the roster.[83]

Meanwhile, similar progress was being made in the effort to replace the antiquated MP54's. The short-haul commuter service in which these cars ran had been operated at a deficit for many years by the railroad. The Pennsylvania could hardly be blamed for spend-

An ignitron tube of the type used in the E-44's. (General Electric Co.)

ing its scarce dollars on improving its electrified freight service, since freight traffic was the railroad's lifeblood. Once it made the decision to stay with electric traction, nonetheless, the PRR had no alternative but to develop a successor to the MP54. Age and deferred maintenance were rapidly taking their toll on the m.u. fleet. In 1958, the Pennsylvania acquired six new stainless steel m.u.'s—Class MP85—from the Budd Company. The MP85's were to serve as prototypes for later production models. Each car could carry 125 passengers at speeds up to 90 miles per hour. In the spring of 1961, the road's Paoli shops removed the four ignitron tubes from one of the MP85's and installed silicon diodes in their place. This was the first instance of silicon rectifiers being used for railway purposes. Successful tests with this equipment paved the way for the subsequent utilization of silicon rectifiers in later m.u.'s and of course in the E-44's.[84]

Owing to the Pennsylvania's poor financial condition, outside help was required before further m.u. cars could be constructed. In 1958, the city of Philadelphia had established the Passenger Service Improvement Corporation (PSIC). Railroad and PSIC engineers cooperated in designing the mass-produced cars which followed the MP85's. In 1963, the railroad placed in service in the Philadelphia area 38 more stainless steel m.u.'s. Dubbed "Silverliners," the new cars were essentially improved versions of the MP85 and were jointly financed by the PRR and PSIC. These Budd-built units featured silicon rectifiers, had a 100-mile-per-hour top speed, and could accelerate at the remarkable rate of 2 miles per hour per second. A few years later, the state of New Jersey purchased an additional 35 Silverliners (slightly improved from the 1963 models) for the railroad to operate in Trenton-New York service.[85]

Monetary assistance from the federal government was made available in 1965 when Congress passed the High Speed Ground Transportation Act. Under this legislation, Washington aided the Pennsylvania in developing a fleet of 160-mile-per-hour cars for service between New York and Washington and in upgrading the roadbed to accommodate this speed. These cars represented a dual attempt to relieve some of the burden on the aging GG1's and to make corridor schedules more competitive with those of the airlines. In 1966, the railroad placed an order for 50 Metroliners, as the cars came to be known, with the Budd Company. Revenue service was expected to begin in 1967, but so many technical flaws appeared during testing that the Metroliners' debut was delayed until January 1969.[86] The troubles that plagued the Metroliners were symptomatic of the status of American railway electrification in the 1960s. The waning of the railroad industry's interest in electric traction since the 1930s caused the technology of electrification to remain relatively static for three

Silverliner No. 201 at Jenkintown, Pa., March 1963, while undergoing trials on Reading Railroad trackage. (Harold K. Vollrath Collection)

decades, with the exceptions previously noted. Now, pressure from the federal government to place the Metroliners in service as soon as possible forced the railroad and its suppliers to compress 10 or 15 years worth of research and development into just 2 or 3. The perils that inevitably accompany such a crash program were well illustrated nearly 40 years earlier by the P5a and the O1; but the development of the Metroliners was destined to teach that lesson anew. Fortunately, once their flaws had been exorcised, the high-speed m.u.'s gave a good account of themselves. The inability to quicken schedules to the extent originally contemplated was not the fault of the Metroliners; rather, it stemmed from the failure of the Pennsylvania and more especially of its private and public successors to improve track and signal systems to permit these faster running times.

A cursory examination of the post-World War II era as a whole would seem to indicate that the PRR entered a period of stagnation with regard to electric traction. Undeniably the period was marked by physical stagnation, despite several studies of the prospect of extending catenary west from Enola, but the railroad had insufficient capital to support such a venture during these years. Electrification had been an extremely expensive proposition even during the 1920s, when the PRR was wealthy. After 1945, prosperity had vanished. The financial subsidies that the federal government began to lavish on competitive forms of transportation (but not on railroads) made a large investment in electric traction even more questionable. Nevertheless, in pioneering the use of the ignitron and silicon-diode rectifiers, the Pennsylvania (along with Westinghouse and General

Metroliner No. 800 at Jenkintown, Pa., October 1967, several months before entering revenue service. (Harold K. Vollrath Collection)

Electric) laid the foundation for more efficient rail transportation in the 1960s and beyond. The two decades since 1945 also witnessed the growing importance of electric traction as a tool for moving freight and the corresponding decline of its role in passenger service. Still, passenger service remained vital, as evidenced by the public sector's growing involvement with electrification in the 1960s. Although the long-distance passenger train was nearing extinction, electrically powered trains in the New York-Washington corridor continued to carry millions of people every year. The assistance of various governmental bodies—though nominal, at first—in bringing about improvements in corridor service and in commuter travel heralded a strong revival of interest in electric traction's unique ability to solve some of the nation's most vexing transportation problems. Just as this new epoch was beginning, however, an old one was drawing to a close. On February 1, 1968, the long-sought merger between the PRR and the New York Central finally reached fruition, and the Pennsylvania Railroad ceased to exist.

Epilogue

....

When the Penn Central Railroad was created, only three long-distance electrifications remained in operation in the United States: the a.c. installations of the Pennsylvania and the New Haven in the Northeast and the Milwaukee Road's high-voltage d.c. system in Montana, Idaho, and Washington. The handful of other roads that had previously used electric traction in heavy-duty, main-line service had gradually phased out the electrics in favor of diesels. Most of these installations comprised but a few route-miles and were originally constructed to overcome special operating problems, such as those presented by long tunnels or steep grades. Diesel locomotives solved these problems nearly as well as electrics and were able to operate beyond the confines of catenary, thus making possible longer runs with a single type of motive power. Diesels eliminated the costly and time-consuming necessity of changing power at either end of the electrification. In short, diesels were so efficient that the high fixed costs associated with the operation of relatively short sections of electrified trackage could no longer be justified.

The Pennsylvania Railroad's electrification successfully withstood the onslaught of the diesel locomotive mainly because of its size. The number of miles under catenary was sufficiently large to enable the economies thus obtained to outdistance fixed costs by a wide margin. Had the railroad electrified only the 35-mile segment through the Allegheny Mountains, as it very nearly did before and after World War I, and had that mileage not been enlarged, electric traction on the PRR would have succumbed to the diesel just as surely as it had on the Norfolk and Western, the Great Northern, and several other roads.

The Pennsylvania was not the only contributor to the Penn Central's electrified lines. The new railroad continued to operate the former New York Central's d.c. electrification at New York, also. Then, shortly after its formation, the Penn Central acquired, by decree of the Interstate Commerce Commission, the New Haven Railroad, which was still operating its a.c.-equipped route between New York and New Haven. In an effort to improve equipment utilization, GG1's soon began running straight through Penn Station to New Haven via the Hell Gate bridge. (This was the new Pennsylvania Station. In what amounted to a shameful act of corporate vandalism, the PRR demolished the original station in 1963, replacing it with a new Madison Square Garden. Much of the trackage beneath the structure remained unchanged, however.) The Long Island Rail Road, on the other hand, a long-time PRR subsidiary, did not figure into the merger, having been taken over by New York's Metropolitan Transportation Authority in 1966.

Contrary to the hopes of its founders, the Penn Central fared far worse than its two major components ever had. In 1971, it declared itself to be bankrupt. Eventually, the federal government took indirect possession of most of the company's property. Amtrak, the quasi-governmental rail passenger corporation established in 1971, assumed ownership of the former Penn Central's electrified lines, with the notable exceptions of the freight route between Enola and Morrisville, and the Columbia and Port Deposit Branch. These lines went to the Consolidated Rail Corporation (Conrail), another government-backed entity encompassing the Penn Central and five other bankrupt northeastern railroads and some of their subsidiaries. In 1977, Amtrak began modernizing its share of the electrification. Plans called for a conversion to 60-cycle, 25,000-volt current, the procurement of much new motive power, the installation of a completely new signal system, and the upgrading of track and structures.[1] Unfortunately, technical problems, last-minute revisions, and bureaucratic infighting among the various public agencies involved have caused work to fall behind schedule.

Conrail, still beset with most of the ills that cursed its predecessors, is debating whether to follow Amtrak's lead and modernize its electrified system or de-energize it entirely and rely exclusively on diesels to power its trains. Should it choose the first alternative, Conrail may well extend catenary across the mountains to Pittsburgh, too, in order to take better advantage of the economies of scale present in electrified operations. (Conrail also maintains trackage rights over many of Amtrak's electrified lines.) In 1976, as part of the legislation creating Conrail, Congress pledged $200 million in loan guarantees in the event the railroad decided to convert its main line between Enola and Pittsburgh to electric traction. If electrically powered trains do begin traversing the Alleghenies someday, a dream more than seven decades old will at last have been fulfilled.[2]

From its inception on the Pennsylvania Railroad, electric traction met with considerable success, first in limited applications at New York and Philadelphia and later when employed to power heavy trains over long distances. The triumph of electrification is best reflected in the marked increase in efficiency that electric motive power brought to passenger and freight operations east of the Susquehanna River, as well as in the longevity of much of its physical plant. To be sure, this longevity was partly owing to the PRR's inability to modernize much of its electrified installation in the post-World War II era; but the fact that electrification was still giving good service even well into the 1970s was a tribute to the superiority of its design and construction. The superiority of electrification was most clearly evident prior to the 1940s, during an era when electricity alone challenged steam. Even in later years, however, when the new diesel technology confronted obsolescent electric equipment, the advantages of electric traction were still sufficient to warrant its retention and eventual improvement (if not expansion) by the PRR.

Precisely how much additional traffic the railroad gained by electrifying is impossible to calculate. What can be stated with assurance is that electrification alone was not able to check the inroads made by competitive forms of transportation. Passenger traffic in particular suffered from this competition, even though the retention and improvement of passenger service had been one of the principal factors underlying the Pennsylvania's decision to electrify. On the other hand, both passenger and freight business might well have declined even more had it not been for the high quality of service permitted through the use of electric traction. Surely the road's financial plight would have worsened in the years following World War II had the PRR not owned an extensive network of electrified trackage.

Other conclusions derived from the Pennsylvania's experience are worth studying by Amtrak, Conrail, and any other railroads that might contemplate initiating or expanding electric operations. The PRR discovered, for example, that the obstacles to large-scale electrification were mainly financial, rather than technological. Undoubtedly, the railroad suffered its share of technical misfortunes, especially with regard to motive power. These difficulties arose primarily because the conservatism of the road's mechanical engineers prevented them from consistently translating their success in designing steam locomotives to the realm of electricity. When the Pennsylvania began to rely more heavily on its consultants and the equipment manufacturers, these technological problems were more readily resolved. Moreover, since the kind of high-speed, long-distance operation the PRR electrification demanded was unlike any other attempted in the United States, the railroad desired a very gradual and methodical approach. In insisting upon exhaustive testing of new equipment before mass production, and in evaluating the electric traction experience (however inadequate) of other roads before committing itself to action, the Pennsylvania unquestionably saved itself a great deal of time and money, while losing relatively little in the way of competitive advantage.

Indeed, a conservative outlook toward electrification was imperative, given the tremendous expense of a long-distance installation. The PRR demonstrated that electrification entailed much more than the mere conversion of motive power. The road realized as early as 1910 that it could not obtain all the advantages inherent in electrified operation until it undertook numerous improvements in such related areas as communication and signal systems, roadbed, and terminal facilities, all of which added to the crushing financial burden that accompanied the adoption of electric traction. The Pennsylvania could not be expected to embark on such a high-priced series of improvements until it had made every effort to keep risks to a minimum. Electrification's financial risk alone was intimidating, and it was one which the railroad could do little to avoid. The PRR did have the capability to control technological risk to a much larger degree, a capability which it usually exercised to the fullest extent.

The high cost of electrification demanded that any railroad management willing to initiate a widespread conversion to electric motive power have faith in the future of rail transport and a firm commitment to seeing the project through to completion, regardless of the hardships encountered. In the Atterbury administration, at least, the Pennsylvania had that kind of management. The application of electric traction on a grand scale required far more foresight and resolve than did technological innovation. In this sense, the PRR was not at all conservative, but rather very progressive.

Cooperation was another essential ingredient underlying the successful application of electric traction. At nearly every level, electrification was very much a joint effort. The Pennsylvania would never have been able to carry out any of its various electrification projects had it not enjoyed the competent assistance of its consultants, equipment suppliers, and the utilities. By their willingness to adapt their operations to meet railroad requirements and by offering electric power at attractive rates, Philadelphia Electric and other utilities played particularly important roles.

Successes notwithstanding, the Pennsylvania also showed that electrification was only one technological tool that a railroad could use to reduce costs and improve service. The PRR was never under any illusions about the potential of electric traction and its accompanying limitations. The use of electric motive power was by no means a panacea. In itself, it could not cure the long-term ills of the Pennsylvania or any other railroad.

Nor did electrification produce a notable savings in the consumption of energy, a subject of much concern in recent times. Electric locomotives did utilize fuel much more effeciently than their steam counterparts, but their advantage over diesel locomotives in this respect is not so pronounced. Electric traction's principal virtue today is its reliance on a plentiful domestic fuel source—coal—rather than on a scarce foreign-based one—oil. Should the price of oil increase more sharply than the cost of electric power (supplied from coal-fired generating stations), conditions will again become favorable for the use of electric motive power by many roads having dense traffic flows.[3] In that case, the Pennsylvania Railroad's experiences with electrification will take on renewed importance.

Notes

••

Abbreviations

AAR	Association of American Railroads
CH	George H. Burgess and Miles C. Kennedy, *Centennial History of the Pennsylvania Railroad Company* (Philadelphia: PRR, 1949).
CR Merion Ave.	Consolidated Rail Corporation, Merion Avenue Records Center, Philadelphia, Pa.
CR Thirtieth St.	Consolidated Rail Corporation, Office of Engineering and Research, Thirtieth Street Station, Philadelphia, Pa.
EJ	*Electric Journal*
ERJ	*Electric Railway Journal*
GE Erie	General Electric Company, Transportation Systems Business Division, Erie, Pa.
IEEE	*Institute of Electrical and Electronics Engineers Transactions*
JFI	*Journal of the Franklin Institute*
NELA	*National Electric Light Association Bulletin*
NYT	*New York Times*
PP I	Alvin F. Staufer and Bert Pennypacker, *Pennsy Power* (Medina, Ohio: Alvin F. Staufer, 1962).
PP II	Alvin F. Staufer and Bert Pennypacker, *Pennsy Power II* (Medina, Ohio: Alvin F. Staufer, 1968).
PRR OEE	Pennsylvania Railroad, Office of the Electrical Engineer
RA	*Railway Age*
RAG	*Railway Age Gazette*
RG	*Railroad Gazette*
SA	*Scientific American*
SR	William D. Middleton, *When the Steam Railroads Electrified* (Milwaukee: Kalmbach Books, 1974).
TAIEE	*Transactions of the American Institute of Electrical Engineers*
TASCE	*Transactions of the American Society of Civil Engineers*
TASME	*Transactions of the American Society of Mechanical Engineers*

Introduction

1. The builder of this locomotive was Charles G. Page. For a detailed account of the experiment, see Robert C. Post, "The Page Locomotive: Federal Sponsorship of Invention in Mid-Nineteenth Century America," *Technology and Culture* 13 (April 1972):140-69. No adequate histories of other early experiments exist. Two works that offer surveys of this early phase of the development of electric traction are Carl Condit, "The Pioneer Stage of Railroad Electrification," *Transactions of the American Philosophical Society* 67 (November 1977):3–5; and *SR*, pp. 11–14.

2. Good summaries of the work of Edison and other notable experimenters in the field of electric traction can be found in Condit, "Pioneer Stage," pp. 5–10; Harold C. Passer, *The Electrical Manufacturers* (Cambridge: Harvard University Press, 1953), pp. 211–36; and Thorburn Reid, "Some Early Traction History," *Cassiers Magazine* 16 (August 1899):357–70.

3. Frank J. Sprague, "Digging in the 'Mines of the Motors,' " *Electrical Engineering* 53 (May 1934):696–702; Passer, *Electrical Manufacturers*, pp. 237–49.

4. The standard history of the interurban lines is by George W. Hilton and John F. Due, *The Electric Interurban Railways in America* (Stanford: Stanford University Press, 1960). See also William J. Clark, "Electric Railways in America from a Business Standpoint," *Cassiers Magazine* 16 (August 1899): 520–21; Joseph Wetzler, "Will the Electric Motor Supersede the Steam Locomotive?" *Scribner's Magazine* 17 (May 1895):594–95.

5. Wetzler, "Electric Motor," pp. 595–600; Hilton and Due, *Interurban Railways,* pp. 22–24, 45–86, 119.

6. For more information on the Nantasket Beach installation, see *RG*, 14 June 1895, pp. 377–78; 2 August 1895, p. 512; and Howard S. Palmer, *The First Steam Railroad Electrification* (n.p.: Newcomen Society of England, 1945). Consult sources in Chapter 3 for the Burlington and Mount Holly.

7. *RG*, 19 July 1895, pp. 480–81; 8 November 1895, pp. 735–36; John Gilmore Speed, "Electricity and Transportation," *Harper's Weekly,* 31 August 1895, pp. 821–22. The trough system proved to be unreliable and was replaced by the more conventional third rail in 1902.

8. Paul T. Warner, "Motive Power Development, Pennsylvania Railroad System," *Baldwin Locomotives* 3 (July 1924):33–57; *CH*, pp. 292–93, 709–20.

9. *CH*, pp. 61–452 passim.

Chapter 1

1. *CH*, pp. 195–219, 241–44.

2. Brian J. Cudahy, *Rails Under the Mighty Hudson* (Brattleboro, Vermont: Stephen Greene Press, 1975), pp. 7–8; *SA*, 22 February 1908, p. 124; *NYT*, 22 July 1880, p. 1; *SA*, 22 February 1908, p. 124.

3. Samuel Rea, "Engineering Reminiscences," *JFI* 202 (August 1926):172–73.

4. Quoted in F. H. Frankland and F. E. Schmitt, "Memoir of Gustav Lindenthal," *TASCE* 105 (1940):1791.

5. Ibid., pp. 1790–94.

6. *The New York Improvements of the Pennsylvania Railroad* (Philadelphia: PRR, 1910), pp. 4–5; Rea, "Reminiscences," p. 173.

7. U.S. Congress, House, Committee on Commerce, *Bridge Across the Hudson River at New York City,* H. Rept. 928, 51st Cong., 1st sess., 1890, *House Reports* 3:19–22; *CH*, pp. 464–65. See also *Bridge Across the Hudson,* p. 1.

8. Committee on Commerce, *Bridge Across the Hudson,* pp. 19–22; *SA*, 23 May 1891, pp. 319, 323; O. H. Ammann, "George Washington Bridge: General Conceptions and Development of Design," *TASCE* 59 (October 1933):8.

9. "Message from the President of the United States," *Congressional Record,* 51st Cong., 1st sess., 1890, 21, pt. 8:7200.

10. Committee on Commerce, *Bridge Across the Hudson,* p. 4.

11. Through freight would not have been affected by a bridge or tunnel into Manhattan, since it would not have been routed through the city anyway. A. J. County, "The Economic Necessity for the Pennsylvania Railroad Extension into New York City," *Annals of the American Academy of Political and Social Science* 29 (March 1907):1.

12. Ibid.; *SA*, 23 May 1891, pp. 319, 323. See also Rea, "Reminiscences," pp. 173–74; Charles W. Raymond, "The New York Extension of the Pennsylvania Railroad," *TASCE* 68 (September 1910):1–3.

13. Rea, "Reminiscences," pp. 165–72.

14. P. V. McMahon, "The City and South London Railway," *Cassiers Magazine* 16 (August 1899):527–40. See also *CH*, pp. 465, 467; Charles M. Jacobs, "The New York Tunnel Extension of the Pennsylvania Railroad: North River Division," *TASCE* 68 (September 1910):35. The September and October 1910 issues of *TASCE* were devoted exclusively to the construction of the tunnel extension and contain a wealth of valuable technical data.

15. *CH*, pp. 465, 467.

16. Ammann, "George Washington Bridge," pp. 2–8.

17. U.S. Congress, Senate, *Report of the Board of Engineers on New York and New Jersey Bridge*, S. Rept. 12, 53rd Cong., 3d sess., 1894, *Senate Executive Documents* 1:37, 41–53; Frankland and Schmitt, "Lindenthal," p. 1791.

18. *Fifty-fifth Annual Report of the Pennsylvania Railroad Company* (Philadelphia: PRR, 1902), p. 21.

19. *CH*, pp. 455–57; *NYT*, 29 December 1906, p. 1.

20. Jacobs, "North River Division," pp. 33–35.

21. *Fifty-foruth Annual Report*, p. 22.

22. Ibid.

23. *CH*, pp. 473–81.

24. *RG*, 15 January 1897, p. 39, *CH*, p. 465.

25. *Railway and Engineering Review*, 11 December 1897, p. 707.

26. *Fifty-fifth Annual Report*, pp. 21–22; *CH*, p. 467.

27. Rea, "Reminiscences," pp. 172–74.

28. Committee on Commerce, *Bridge Across the Hudson*, p. 36.

29. *Fifty-fifth Annual Report*, p. 22; *RAG*, 30 November 1917, p. 974.

30. *NYT*, 12 December 1901, p. 1; Committee on Commerce, *Bridge Across the Hudson*, pp. 19–22; Cassatt: quoted in the *NYT*, 12 December 1901, p. 1.

31. *Fifty-fifth Annual Report*, p. 22.

32. Raymond, "New York Extension," pp. 3–8.

33. *Fifty-sixth Annual Report*, 1903, p. 19.

34. *Fifty-sixth Annual Report*, p. 22.

35. Lindenthal: Frankland and Schmitt, "Lindenthal," pp. 1790–94. The George Washington Bridge, completed in 1933, should be considered only a partial realization of Lindenthal's dream, since it makes no provision for rail mass transit; Noble: Ralph Modjeski, Onward Bates, and Isham Randolph, "Memoir of Alfred Noble," *TASCE* 79 (December 1915): 1352–74; Brown: *CH*, p. 465.

36. Jacobs, "North River Division," p. 40; Gilbert H. Gilbert, Lucius I. Wightman, and William L. Saunders, *The Subways and Tunnels of New York: Methods and Costs* (New York: Wiley, 1912), pp. v, 5.

37. Quoted in R. W. Raymond and Alfred Noble, "Memoir of Charles Walker Raymond," *TASCE* 77 (December 1914):1894–1901.

38. Raymond, "New York Extension," pp. 17, 19–20.

39. Ibid., p. 17.

40. *Sixty-first Annual Report*, 1908, p. 20; *Fifty-sixth Annual Report*, p. 19.

41. Raymond, "New York Extension," pp. 17–18.

42. Ibid., pp. 23–24; *SA*, 9 August 1890, pp. 87–88; 13 September 1890, p. 164.

43. Raymond, "New York Extension," pp. 17–18; George Gibbs, "The New York Tunnel Extension of the Pennsylvania Railroad: Station Construction, Road, Track, Yard Equipment, Electric Traction, and Locomotives," *TASCE* 68 (October 1910):247–48, 238–55; George C. Clarke, "The New York Tunnel Extension of the Pennsylvania Railroad: The Site of the Terminal Station," *TASCE* 68 (October 1910):363–88.

44. E. B. Temple, "The New York Tunnel Extension.of the Pennsylvania Railroad: Meadows Division and Harrison Transfer," *TASCE* 68 (September 1910):74–89.

45. *NYT*, 3 April 1910, p. 20; Cudahy, *Rails Under*, p. 29. Cataracts had caused Colonel Raymond to suffer nearly total blindness by 1904, yet Cassatt would not hear of dismissing him. Indeed, throughout the period of construction, Raymond continued to provide a strong guiding hand and often involved himself in the most intricate engineering details. Raymond and Noble, "Memoir of Raymond," pp. 1894–1901.

46. *SA*, 20 October 1906, p. 278; 22 February 1908, p. 126; Raymond, "New York Extension," pp. 25–28.

47. *SA*, 1 November 1890, pp. 280–81; 22 February 1908, p. 124.

48. William Gibbs McAdoo, *The Crowded Years* (Boston: Houghton Mifflin, 1931), pp. 68, 70–79.

49. *NYT*, 12 December 1901, p. 1; Jacobs, "North River Division," p. 41.

50. McAdoo, *Crowded Years*, p. 86; *RG*, 21 February 1908, pp. 241–47.

51. McAdoo, *Crowded Years*, pp. 89–90.

52. Ibid., p. 91; Cassatt: quoted in ibid., p. 92.

53. McAdoo, *Crowded Years*, p. 93; Cudahy, *Rails Under*, pp. 19–21, 50. The PRR discontinued ferry service to and from its Exchange Place Station in 1950 and abandoned the station altogether in 1961.

54. Temple, "Meadows Division," pp. 75–78.

55. William J. Wilgus, "The Electrification of the Suburban Zone of the New York Central and Hudson River Railroad in the Vicinity of New York City," *TASCE* 61 (December 1908):73–74; *NYT*, 9 January 1902, p. 1.

56. Wilgus, "New York Central Electrification," pp. 73–74; *NYT*, 24 January 1902, p. 14; 25 January, p. 2; 31 January, p. 2; 21 November 1903, p. 3; 22 November, p. 13.

57. Wilgus, "New York Central Electrification," pp. 73–74, 87–89, 94–95; *Railway Gazette*, 13 February 1913, p. 279.

58. E. H. McHenry, "Electrification of the New York, New Haven, and Hartford," *RG* 43 (16 August 1907):177–84.

59. H. C. Griffith, "Single-Phase Electrification on the Pennsylvania Railroad," *Journal of the Institution of Electrical Engineers* 81 (July 1937):91; H. A. Dahl, "Development of Electric Traction on the Pennsylvania Railroad," *RA* 87 (6 July 1929):4.

60. "Electrification of the Long Island Railroad," *RG* 39 (5 November 1905):412–15; O. S. Lyford, "The Electrification of the Long Island Railroad," *EJ* 3 (January 1906):29.

61. *SR*, pp. 268, 430–31.

62. *CH*, p. 559; County, "Economic Necessity," pp. 6–7. ·

63. *Fifty-fourth Annual Report*, p. 22; *New York Improvements*, p. 708.

64. *Fifty-fourth Annual Report*, p. 22.

65. Raymond, "New York Extension," pp. 3–4.

66. Ibid.

67. "Electrification of the New York Connecting Railroad," *RA* 64 (7 June 1918):1367–70.

Chapter 2

1. George Gibbs, "Station Construction," pp. 329–30.

2. Warner, "Motive Power Development" (July 1924):33–57; (October 1924):3–9; *PP* I, 6–7, 10–39.

3. *RA*, 1 June 1940, p. 996; David B. Sloan, *George Gibbs and E. Rowland Hill: Pioneers in Railroad Electrification* (New York: Newcomen Society of North America, 1957), pp. 8–12; *RG*, 29 September 1905, p. 301; McHenry, "Electrification of the New Haven," p. 177.

4. Wilgus, "New York Central Electrification," p. 105; George Gibbs, "Station Construction," p. 330.

5. Ibid., pp. 129–38.

6. Ibid., pp. 139–40, 164–75, 329–34.

7. Frank J. Sprague, "Past and Current Developments of Electric Traction," *NELA* 19 (August 1932):461–66.

8. N. W. Storer, "Characteristics of Electric Locomotives," *JFI* 192 (October 1921):459–60; Arthur J. Manson, *Railroad Electrification and the Electric Locomotive* (New York: Simmons-Boardman, 1925), pp. 36–37, 141.

9. Manson, *Railroad Electrification*, p. 142; George Westinghouse, "The Electrification of Railways," *TASME* 32 (1910):956.

10. William Bancroft Potter, "The Economies of Railway Electrification," *TASME* 32 (1910):911–14; Storer, "Characteristics," p. 463. See also Potter, "Economies," pp. 915–16.

11. Sprague, "Motors," pp. 705–6. The New York Central steam locomotive was a new K-class 4-6-2. *RG*, 26 May 1905, pp. 584–86; *SA*, 18 February 1905, p. 142.

12. Wilgus, "New York Central Electrification," pp. 84, 94. The Class S designation resulted from the addition of an extra guiding axle to the Class T machines, a modification the Central hoped would improve the locomotives' tracking qualities.

13. Ibid., pp. 96–102. Third rail was an energized rail situated adjacent and parallel to the running rails. Current collection was accomplished by means of a contact shoe extending from the locomotive or car. Third-rail systems, unlike catenary systems, were best suited for low-voltage, direct current installations.

14. *RG*, 29 September 1905, p. 301; McHenry: "Electrification of the New Haven," pp.177–81.

15. "Single-Phase Locomotive for Heavy Railroad Service," *Street Railway Journal* 25 (3 June 1905):999–1001.

16. McHenry, "Electrification of the New Haven," pp. 181–82.

17. Contact wire was the energized wire in the catenary system that delivered the electric current directly to the pantograph shoe of a locomotive.

18. Pantograph was a device mounted atop a car or locomotive to facilitate current collection from the contact wire. The pantograph consisted of a hinged, framelike-arm that could be raised or lowered. When raised, its uppermost part (the shoe) touched the contact wire and completed the circuit needed for current collection. The pantograph was much stronger and better suited for high-speed service than the single-arm trolley pole, which it replaced shortly after the turn of the century.

19. W. S. Murray, "The Log of the New Haven Electrification," *TAIEE* 27-2 (11 December 1908):1618–24.

20. Griffith, "Single-Phase," p. 91; George Gibbs, "Station Construction," pp. 330–31.

21. Lyford, "Electrification of the Long Island," pp. 29–32.

22. *RA*, 1 June 1940, p. 996.

23. George Gibbs, "Station Construction," p. 363.

24. *RG*, 30 September 1907, pp. 327–28; Manson, *Railroad Electrification*, pp. 217–19; Frederick Westing, "The Locomotive That Made Penn Station Possible," *Trains* 16 (October 1956):30.

25. George Gibbs, "Station Construction," pp. 358–59.

26. Alfred W. Gibbs, "Some Mechanical Characteristics of High Speed, High Power Locomotives," *JFI* 192 (October 1921):469.

27. "The Electrical Equipment of the West Jersey and Seashore Branch of the Pennsylvania Railroad," *Street Railway Journal* 28 (10 November 1906):928–46.

28. Alfred W. Gibbs, "High Speed Locomotives," pp. 476–79.

29. *RG*, 20 September 1907, pp. 327–28.

30. George Gibbs, "Station Construction," pp. 361–62.

31. Alfred W. Gibbs, "High Speed Locomotives," pp. 474–75; *RG*, 22 November 1907, pp. 624–26.

32. *NYT*, 13 July 1907, p. 5.

33. PRR D16b locomotives included Nos. 6034 and 6047. Class E2's included Nos. 6020, 6069, and 6075. The EP-1 was No. 028. Alfred Gibbs, "High Speed Locomotives," p. 491; *NYT*, 14 November 1907, p. 1. See also Alfred Gibbs, "High Speed Locomotives," p. 483.

34. Onlookers' observations: *Philadelphia Inquirer*, 17 November 1907, p. 5; *NYT*, 17

November 1907, p. 4; Franklinville tests: "Experimental Overhead Trolley Construction of the Pennsylvania Tunnel and Terminal Railroad," *ERJ* 32 (12 December 1908):1546.

35. Alfred Gibbs, "High Speed Locomotives," pp. 491–92.
36. "Experimental Overhead Trolley," pp. 1546–52.
37. George Gibbs, "Station Construction," p. 331.
38. *ERJ*, 12 December 1908, p. 1542.
39. A. Frederick Collins, "The Long Island Railroad Electrified," *SA* 93 (12 November 1905):397–98; George Gibbs, "Station Construction," p. 315.
40. Murray, "Log of the New Haven," p. 1663.
41. "Discussion of a Paper Presented by W. S. Murray," *JFI* 180 (July 1915):76.
42. Alfred Gibbs, "High Speed Locomotives," pp. 492–94.
43. Ibid., pp. 473–75; George Gibbs, "Station Construction," p. 361.
44. Alfred Gibbs, "High Speed Locomotives," pp. 475–77; Manson, *Railroad Electrification*, pp. 224–28.
45. *NYT*, 3 November 1908, p. 11; *RG*, 7 May 1909, p. 999; *NYT*, 2 May 1909, p. 2.
46. George Gibbs, "Station Construction," p. 363; Griffith, "Single-Phase," pp. 99–100; "Pennsylvania Electric Locomotives," *Railroad Age Gazette* 47 (5 November 1909):881–84.
47. *Railway and Engineering Review*, 4 March 1910, p. 172.
48. *NYT*, 2 August 1910, p. 7.
49. Ibid., 9 September 1910, p. 1; *RAG*, 2 December 1910, p. 1087.
50. *New York Improvements*, p. 30.
51. F. H. Shephard, "The Greatest Railroad Work in History," *EJ* 8 (January 1911):29.
52. H. L. Kirker, "The Pennsylvania Electric Locomotives and Their Field of Operation," *EJ* 7 (September 1910):668.
53. *SA*, 18 December 1909, p. 464.
54. George Gibbs, "Station Construction," p. 232; *PP I*, p. 249.
55. George Gibbs, "Station Construction," p. 232; *RAG*, 3 March 1911, p. 439.
56. "New York Connecting Railroad," pp. 1367–70.
57. "Electric Equipment," in *Proceedings of the Forty-sixth Annual Convention of the American Railway Master Mechanics Association* (Chicago: Henry O. Shephard, 1913), pp. 131–32; "Maintenance of the Pennsylvania Railroad Electric Locomotives," *ERJ* 41 (15 March 1913): 452–61.
58. Accumulated mileage and failures: *RAG*, 17 September 1915, p. 513; Homer K. Smith, "Some Service Records of Electric Equipment," *RA* 73 (25 November 1922):1013; C. C. Whittaker, "Electrification Reviewed from the Standpoint of Service," *Railway Review* 68 (19 March 1921):487.
59. "Westinghouse Electric and Manufacturing Company Exhibit at the Panama-Pacific Exposition," *EJ* 12 (June 1915):19–22. The PRR attached separate road numbers to each DD1 unit, while giving an additional electrified zone number to each two-unit locomotive. The DD1 on exhibit in San Francisco, for example, carried road Nos. 3938 and 3939 and electrified zone No. 36.
60. M. C. Turpin, "A Remarkable Exhibit of Railway Apparatus," *EJ* 12 (October 1915):477–82.

Chapter 3

1. *Sixty-fifth Annual Report*, 1912, p. 10.
2. *CH*, pp. 431–33, 497–98.
3. W. T. Whalen, "Electrification Solves the Congested Traffic Problem in Philadelphia," *Railway Review* 66 (January 1920):86–87; Smith, "Service Records," p. 104.
4. Nathaniel Burt, *The Perennial Philadelphians* (Boston: Little, Brown, 1963), pp. 196–97. "Main Line" originally referred not to the PRR's tracks, but to the older "main line of public works," a series of canals and rail lines that the state had constructed early in the nineteenth century to give Philadelphia an improved link with

the west. The PRR purchased most of these works, however, and its right-of-way followed their approximate course.

5. *Electrical Engineer*, 12 June 1895, p. 535; *SR*, pp. 23, 25.

6. *Electrical Engineer*, 17 April 1895, p. 347; 14 August 1895, pp. 156–57. The PRR specified that the first two cars be delivered with two motors of 75 horsepower each, while the third car should have four 50-horsepower motors.

7. Electrification of other branch lines: *Philadelphia Inquirer*, 23 July 1895, p. 2; *RG*, 26 July 1895, p. 504; Jackson and Sharp cars: *Electrical Engineer*, 12 June 1895, p. 535.

8. *RA*, 5 April 1895, pp. 161–62.

9. *Electrical Engineer*, 14 August 1895, pp. 156–57.

10. *NYT*, 28 November 1901, p. 3; "Electrification," *Pennsy* 3 (July–August 1954):13.

11. Sloan, *Gibbs and Hill*, pp. 8–26.

12. George Gibbs, "The Philadelphia-Paoli Electrification of the Pennsylvania Railroad," *EJ* 13 (February 1916):68–78.

13. Ibid., p. 70; *PP* II, pp. 168–70.

14. George Gibbs, "Station Construction," p. 331.

15. Electrification project: see *ERJ*, 12 December 1908, pp. 1543–44. Maintenance of Equipment Committee on Power Plants of the Pennsylvania Railroad, *Electric Traction—In Its Present Relation to Steam Railroads and Its Possible Use with Reference to Hauling of Heavy Trains* (Altoona, Pa.: PRR, 1909), pp. 5–32. See also George Gibbs, "Philadelphia-Paoli," pp. 70–71; *Railway and Engineering Review*, 15 March 1913, p. 219.

16. *Sixty-seventh Annual Report*, 1914, p. 11; "Electrification," p. 14.

17. The exception was the Michigan Central's Detroit River tunnel line, opened in 1910 and comprising 4.5 route-miles. The railroad contracted with the Detroit Edison Company for power. Fred Darlington, "Central Station Power Plants and Electricity Supply for Trunk Line Railroads," in *Proceedings of the Thirty-fourth Convention of the National Electric Light Association* (New York:NELA, 1911), pp.1064–65; *SR*, pp. 142–48.

18. Federal Power Commission, National Power Survey, *The Use of Electric Power in Transportation*, Power Series No. 4 (Washington: Government Printing Office, 1936), p. 32; E. P. Dillon, "Electric Railway Loads on Central Stations," in *Proceedings of the Thirty-sixth Convention of the National Light Association* (New York: NELA, 1913), p. 698.

19. Darlington, "Central Station Plants," pp. 1064–65.

20. W. C. L. Eglin, "Purchased Power for Electric Railways," *EJ* 11 (October 1914):504; *Electrical World* 15 (July 1916): 113.

21. Darlington, "Central Station Plants," pp.1064–65; Horace Liverslidge, *Electric Service in Philadelphia Since 1881* (New York: Newcomen Society of England, 1945), p. 21.

22. Liverslidge, *Philadelphia*, p. 21; W. C. L. Eglin, "The Engineering Features of the Philadelphia Electric Company System," *Electrical World* 83 (10 May 1924):933.

23. Eglin, "Engineering Features," pp. 934–45; Nicholas B. Wainwright, *History of the Philadelphia Electric Company* (Philadelphia: Philadelphia Electric, 1961), pp. 107–9.

24. Formal talks: Wainwright, *Philadelphia Electric*, pp. 110, 105–10; plant expansion: *Electrical World*, 3 January 1914, p. 26.

25. *Electrical World*, 3 January 1914, p. 26; *NYT*, 25 September 1913, p. 15; *Sixty-seventh Annual Report*, p. 11.

26. *RA*, 15 July 1922, p. 120; Eglin, "Purchased Power," p. 504; Wainwright, *Philadelphia Electric*, p. 111.

27. George Gibbs, "Philadelphia-Paoli," pp. 71–75; "Electrification of the Pennsylvania at Philadelphia," *RAG* 59 (12 November 1915):891.

28. The New York tunnel trackage employed a system of colored lights exclusively and was the only other segment of PRR main line not to use semaphores at this time. "Electrifying One of the Busiest Stretches of Railroad in America," *PRR Information for the Employees and the Public* 3 (1915):8–11.

29. *Railway World*, April 1915, pp. 23–31; *Railway Review*, 19 June 1915, p. 844.

30. Westinghouse, "Electrification of Railways," p. 958; Dillon "Electric Railway Loads," p. 706.

31. *SR*, pp. 115, 414–15.

32. E. B. Shew and W. E. Kelley, "Railroad Electrification in the United States," in *Proceedings of the American Power Conference* (Chicago: Illinois Institute of Technology, 1967), p. 889.

33. Liverslidge, *Philadelphia,* pp. 29–30; Wainwright, *Philadelphia Electric,* p. 111.

34. Christian Street Station: *Electrical World,* 9 October, 1915, p. 800; power director: "Pennsylvania Electrification at Philadelphia," *Electrical World* 66 (13 November 1915):1074–75.

35. Of the 93 MP-54's originally built, 82 were passenger cars, 9 were passenger-baggage combinations, and 2 were baggage-mail combines. George Gibbs, "Philadelphia-Paoli," p. 70; "Electrification of the Pennsylvania at Philadelphia," pp. 892–93.

36. *Philadelphia Inquirer,* 11 September 1915, p. 2; 12 September 1915, p. 4. See also W. H. Thompson and L. E. Frost, "Operation of the Philadelphia-Paoli Electrification of the Pennsylvania Railroad," *EJ* 13 (October 1916):485.

37. Laurence M. Willson, "Operating Experiences on the Philadelphia-Paoli Electrification," *EJ* 14 (October 1917):396-98. Class D 4–4–0's, long since superseded in main-line duties by more modern steam power, had been handling most of the suburban traffic in the Philadelphia area prior to the advent of electrification. The elderly 4–4–0's soon gave way on nonelectrified routes to larger Class G 4–6–0's. See also Homer Smith, "Service Records," p. 1014.

38. "Busiest Stretch of Railroad," p. 1; "Philadelphia-Chestnut Hill Electrification," *RA* 64 (10 May 1918):1177–81; "P. R. R. Extends Electrification to Chestnut Hill," *ERJ* 51 (27 April 1918):798–803.

39. "Busiest Stretch of Railroad," p. 7; "Electrification of the Pennsylvania at Philadelphia," pp. 892–93. The railroad also equipped its fleet of DD1's with these "dead man's" controls, although, unlike the MP54's, the DD1's always carried two men in the cab. If the engineman became incapacitated, therefore, his helper could have quickly taken the controls. The addition of the "dead man's" device was probably aimed at allaying the fears of the public in that case. In regard to the m.u. cars, however, such safety precautions satisfied a real need.

40. "Electrification," p. 14.

41. H. A. Dahl, "Development of Electric Traction on the Pennsylvania, *RA* 87 (6 July 1929):6.

42. F. C. Grimshaw, "Operation of the Philadelphia-Paoli Electrification," *ERJ* 47 (8 April 1916):683; Willson, "Operating Experiences," p. 396.

43. "Electrification," p. 14; "Danger: Live Wire, Keep Off," *Pennsy* 2 (August 1953):12.

44. George Gibbs, "Philadelphia-Paoli," p. 68; "Electrification of the Pennsylvania at Philadelphia," p. 889; Whalen, "Congested Traffic Problem," pp. 86–87.

45. George Gibbs, "Philadelphia-Paoli," p. 68.

Chapter 4

1. W. S. Murray, "Conditions Affecting the Success of Mainline Electrification," *JFI* 179 (May 1915):515–18.

2. *CH,* pp. 265–66, 415.

3. Warner, "Motive Power Development" (October 1924):3–4.

4. *Fifty-ninth Annual Report,* 1905, p. 23; Griffith, "Single-Phase," p. 91; *Street Railway Journal,* 10 November 1906, p. 927.

5. *Railway and Engineering Review,* 14 October 1905, p. 729; *Street Railway Journal,* 10 November 1906, p. 927.

6. *RG,* 22 December 1905, p. 594.

7. The PRR originally equipped the Newfield-Millville segment with overhead wire, but replaced it in 1910 with third rail, owing to the much lower maintenence cost of third rail. J. V. B. Duer, "Third Rail and Trolley System of the West Jersey and Seashore Railroad," *TAIEE* 34-2 (1 July 1915):1517–24; *RG,* 6 June 1906, p. 560.

8. "The Electrical Equipment of the West Jersey and Seashore Branch of the

Pennsylvania Railroad," *Street Railway Journal* 28 (10 November 1906):939–40; "Electrification of the West Jersey and Seashore," *RG* 41 (9 November 1906):415.

9. B. F. Wood, "Electrical Operation of the West Jersey and Seashore Railroad," *TAIEE* 30–2 (28 June 1911): 1381, 1378–79; *RG*, 21 September 1906, p. 72.

10. Wood, "Electrical Operation," pp. 1382–83; William J. Clark, "Electrification of Mainline Railroads," *JFI* 173 (June 1912):599–600. The prosperity of the West Jersey and Seashore's electrified service lasted only until the 1920s, when bus and auto competition began to make significant inroads into the road's Philadelphia-Atlantic City business. Electric operations ceased on that route in 1931. The WJ&S and successor Pennsylvania-Reading Seashore Lines continued to operate local m.u. service out of Camden until 1949. *RA*, 10 October 1931, p. 563; *Philadelphia Inquirer*, 25 September 1949, p. 4B.

11. *Fifty-ninth Annual Report*, p. 22; *Fifty-sixth Annual Report*, p. 23.

12. Ibid., p. 24; *CH*, pp. 491–92.

13. New Portage Branch: The New Portage Branch followed the right-of-way of the old state-owned New Portage Railroad, which was part of the famous Pennsylvania Canal system and was abandoned in 1858. In connection with the double-tracking of this branch in 1903, the PRR also double-tracked the Petersburg Branch. *Fifty-sixth Annual Report*, p. 24. See also survey: *Fifty-ninth Annual Report*, p. 22.

14. *Sixty-second Annual Report*, 1908, p. 9.

15. *Fifty-ninth Annual Report*, p. 22.

16. H. C. Griffith, "Electric Locomotive Operation," *RA* 111 (9 August 1941):230–31; Manson, *Railroad Electrification*, pp. 75–88; "Evolution from Steam to Electric Traction," *RA* 107 (23 September 1939):443.

17. Griffith, "Electric Locomotive Operation," pp. 230–31; Manson, *Railroad Electrification*, pp. 75–88.

18. H. L. Andrews, "The Electric Locomotive Improves Its Capability," *RA* 90 (24 January 1931):242.

19. Maintenance of Equipment Committee, *Electric Traction*, pp. 18–31.

20. Ibid., pp. 31–32.

21. *SR*, pp. 155–65, 428–35.

22. Maintenance of Equipment Committee, *Electric Traction*, pp. 26–38.

23. *Sixty-seventh Annual Report*, pp. 12–13.

24. *Sixty-ninth Annual Report*, 1916, p. 9.

25. Ibid., pp. 9–21.

26. *Seventieth Annual Report*, 1917, p. 10.

27. G. M. Eaton and A. J. Hall, "The New Split-Phase Locomotive of the Pennsylvania Railroad," *EJ* 14 (October 1917):406–12; "Novel Locomotive for the Pennsylvania," *ERJ* 49 (9 June 1917): 1048.

28. *Railway Review*, 22 June 1917, p. 870; *NYT*, 8 July 1917, p. 4.

29. The LC-1's and the FF-1 incorporated a gear and side-rod drive in place of the older side-rod mechanism used by the DD1's. The motors were coupled to jackshafts via intermediate gearing rather than through connecting rods, an arrangement that permitted the use of higher speed motors. *RAG*, 8 June 1917, pp. 1199–1200; "Norfolk and Western Elkhorn Grade Electrification," *RAG* 58 (4 June 1915): 1153–63.

30. *Railway Review*, 23 June 1917, p. 870; *RAG*, 5 October 1917, p. 606.

31. Mallets:*RAG*, 1 March 1912, pp. 377–78; Warner, "Motive Power Development" (October 1924): 19–20; *PP* 1, p. 19. See also doubleheading: "Pennsylvania Mikado and Pacific-Type Locomotives," *RAG* 57 (3 July 1914):12–13; *PP* I, p. 51; 2–10–0: "Pennsylvania Locomotive of the Decapod Type," *RAG* 62 (15 June 1917):1241–43; Warner, "Motive Power Development" (October 1924):16–18, 29.

32. "Mikado and Pacific-Type," pp. 12–16.

33. McAdoo, *Crowded Years*, p. 477. McAdoo also served concurrently as secretary of the treasury in President Wilson's administration. In the 1920s, the Georgia-born lawyer became a prominent contender for the Democratic presidential nomination.

34. *Seventy-fourth Annual Report*, 1921, p. 2; *Seventy-fifth Annual Report*, 1922, pp. 4–5.

35. *Seventy-fifth Annual Report*, p. 2; freight trains: see Frederick Westing, "K4s," *Trains* 16 (August 1956):44–48; Richard D. Adams, "Ils," *Keystone* 19 (March 1977):2; Mallet: "Simple Mallet with Short Maximum Cut-Off," *RA* 66 (23 June 1919):1675–81.

36. Quoted in the *NYT*, 10 November 1923, p. 20; See also the *Pittsburgh Press,* 9 November 1923, p. 1.

37. J. V. B. Duer, "The Pennsylvania Railroad Electrification," *TAIEE* 50 (March 1931):102; T. C. Wurts, "Pennsylvania Builds Three Electric Locomotives," *RA*, 76 (26 January 1924):295. The a.c. L5 was rated at 3,070 horsepower, the d.c. L5 at 3,370 horsepower. See Griffith, "Single-Phase," p. 102.

38. Homer K. Smith, "The Virginian Railway Electrification," *RA* 76 (7 June 1924):1153–54; *SR*, pp. 181–89.

39. Discussion of a.c. vs. d.c. controversy appended to Clark, "Mainline Railways," p. 604. For typical discussions of the respective merits of steam and electricity, see "Advantages of Steam and Electric Locomotives," *RA* 69 (29 October 1920):739–46; Arthur Curran, "The Advantages of Steam Over Electricity on Railroads," *Cassiers Magazine* 41 (March 1912):222–34; and Storer, "Characteristics," pp. 453–68.

40. Discussion appended to John E. Muhlfield, "Scientific Development of the Steam Locomotive," *TASME* 41 (December 1919):1052.

41. *Electrical World,* 30 October 1920, pp. 891–92; "Advantages of Steam and Electric Locomotives," pp. 739–46.

Chapter 5

1. *Sixty-fifth Annual Report*, pp. 8–9; *NYT*, 3 April 1931, p. 36.

2. *Sixty-eighth Annual Report*, 1915, p. 11; *Seventeenth Annual Report*, p. 11.

3. *PP* I, pp. 51–53, 65–66, 159–61; Westing, "K4s," pp. 44–48; Adams, "Ils," pp. 2–7.

4. W. S. Murray et al., *A Superpower System for the Region Between Boston and Washington,* United States Geological Survey Paper No. 123 (Washington: Government Printing Office, 1921), p. 9; *NYT*, 7 December 1921, p. 19.

5. Murray et al., *Superpower*, p. 9; *RA*, 18 March 1921, p. 727.

6. Murray et al., *Superpower*, p. 12; *NYT*, 6 November 1921, p. 23.

7. Murray et al., *Superpower*, pp. 12, 50–51, 75–83.

8. Ibid., pp. 64–65, 71.

9. *PP* I, pp. 33-35, 125–26, 144, 159–61, 249.

10. J. V. B. Duer, "What Electric Operation Is Doing," *RA* 116 (8 April 1944):687.

11. T. C. Wurts, "Tentative Standard Electric Locomotive, P. R. R.," *Railway Review* 74 (26 January 1924):170; J. V. B. Duer, "Electric Locomotives for the Pennsylvania," *Electrical Engineering* 51 (July 1932):488.

12. Wurts, "Pennsylvania Builds Three," p. 295; W. H. Eunson, "Pennsylvania R. R. 2–8–2 Type Electric Locomotive," *Railway Review* 74 (21 June 1924):1189–91.

13 Wurts, "Pennsylvania Builds Three," p. 295. See also *PP* I, p. 256.

14. Wurts, "Pennsylvania Builds Three," p. 295.

15. James T. Wallis to Elisha Lee, 4 January 1926, Package No. 4411, CR Merion Ave.

16. James T. Wallis to Elisha Lee, 16 March 1926, Package No. 4411, CR Merion Ave. See also *NYT*, 4 October 1926, p. 39.

17. At least one of the original pair of d.c.-powered L5's was also given high hoods, although, as in the case of all the d.c.-equipped L5's, it never underwent actual modification to alternating current. James T. Wallis to Elisha Lee, 16 March 1926, package No. 4411, CR Merion Ave.; *PP* I, p. 256; Frederick Westing, *Penn Station: Its Tunnels and Siderodders* (Seattle: Superior Publishing, 1978), pp. 164–68. A complete roster of PRR electric locomotives is included in Dan Dover et al., "Roster of Penn Central + NYC + PRR + NH," *Extra 2200 South* 7 (June 1969):18, but it should be used cautiously, since it contains a number of errors.

18. *RA*, 3 May 1924, p. 1080.

19. *Seventy-seventh Annual Report*, 1924, 1925, p. 5. *Seventy-eighth Annual Report,* p. 5. See also *Seventy-ninth Annual Report*, 1926, p. 6; *Railway Review*, 24 July 1926, p. 122.

20. Duer, "What Electric Operation Is Doing," p. 685.

21. Wainwright, *Philadelphia Electric*, pp. 150–52, 190–92.

22. Ibid., pp. 166–72, 178; *Electrical World*, 14 June 1924, p. 1249. Conowingo transaction: *Philadelphia Inquirer*, 14 June 1924, p. 2; *Electrical World*, 14 June 1924, p. 1249.

23. *NYT*, 18 June 1924, p. 28; *Philadelphia Inquirer*, 7 January 1925, p. 1; *NYT*, 7 January 1925, p. 2.

24. Wainwright, *Philadelphia Electric*, pp. 170–71.

25. *Electrical World*, 10 May 1924, p. 926.

26. "A Railroad Executive's View: An Interview with Elisha Lee," *Electrical World* 92 (8 December 1928):1137. See also Allen M. Perry, "The Significance of the Pennsylvania Electrification," *Electrical World* 92 (8 December 1928):1133.

27. *Philadelphia Inquirer*, 11 February 1927, p. 1; *NYT*, 11 February 1927, p. 15; *Electrical World*, 19 February 1927, pp. 416–17.

28. Johnson: quoted in the *Philadelphia Inquirer*, 11 February 1927, p. 1. See also Wainwright, *Philadelphia Electric*, p. 192. Thompson: *Philadelphia Inquirer*, 12 February 1927, p. 1.

29. *Philadelphia Inquirer*, 12 February 1927, p. 1; Wainwright, *Philadelphia Electric*, p. 206.

30. Wainwright, *Philadelphia Electric*, pp. 206–11.

31. *Electrical World*, 30 July 1927, p. 231; *RA*, 6 August 1927, p. 271.

32. Griffith, "Single-Phase," p. 92; "Railroad Executive's View," p. 1136: *CH*, p. 615.

33. *Eighty-first Annual Report*, 1928, p. 9; *One Hundred Fourth Annual Statistical Statement of the Pennsylvania Railroad Company* (Philadelphia: PRR, 1951), pp. 31–32. See also *Eighty-first Annual Report*, p. 2.

34. *CH*, pp. 589–91, 599–604, 634; *NYT*, 22 December 1929, Sec. 9, p. 6; 21 September 1935, pp. 1–2.

35. *CH*, pp. 599–604; *NYT*, 25 April 1935, p. 35.

36. "Pennsylvania Railroad," *Fortune* 13 (June 1936):89–94, 138, 140, 142, 144. See also *One Hundred Fourth Annual Statistical Statement*, p. 31.

37. *Eighty-second Annual Report*, 1929, pp. 7–8; *CH*, pp. 582–87, 626–28.

38. *NYT*, 2 November 1928, p. 38.

39. Duer, "PRR Electrification," p. 102; *PP* I, p. 256; Westing, *Penn Station*, pp. 168–71.

40. D. R. MacLeod, "Recommendations on Electric Freight Locomotives Made to the PRR by GE," 25 August 1948, GE Erie; J. V. B. Duer, "Pennsylvania Develops Three Types of Electric Locomotives," *RA* 92 (21 May 1932): 870–71.

41. *RA*, 3 November 1928, p. 870.

42. Ibid.

43. Ibid.

44. Ibid.

45. Ibid.; J. V. B. Duer, "Power Supply for the Pennsylvania Railroad," *Electrical Engineering* 52 (February 1933):111–15. See also *NYT*, 1 November 1928, p. 1.

46. *RA*, 10 November 1928, p. 914.

47. *Electrical World*, 17 November 1928, p. 976; Samuel Insull, "Some Comments on the Economics of Electricity Supply," *NELA* 13 (June 1926):357.

48. *Philadelphia Inquirer*, 2 November 1928, p. 7.

49. *NYT*, 2 November 1928, p. 24.

50. Ibid.

51. *Commercial and Financial Chronicle*, 2 March 1929, pp. 1283–84.

52. Ibid.

53. *Locomotive Engineers Journal* 62 (December 1928):887.

54. Ibid., February 1932, p. 85.

55. *RA*, 4 August 1928, p. 232; *Electrical World*, 28 March 1931, p. 574; "Benefits That Warranted Pennsylvania Electrification," *Electrical World* 97 (25 April 1931):769.

56. *Coal Age*, 20 November 1924, p. 709.

57. "Benefits That Warranted," pp. 769–70.

58. *Railway Engineering and Maintenance* 28 (August 1932):497. The PRR had briefly considered the idea of going ahead with the construction of the relief line in spite of electrification, since some doubts still persisted concerning even a four-track

electrified line's ability to meet future traffic demands. The relief line would also eliminate the smoke nuisance around Trenton. The War Department cited the unfinished bridge across the Delaware as a hazard to navigation and ordered the railroad to complete its construction or dismantle it. The PRR chose to dismantle it, subsequently terminating plans for a relief line once and for all. *NYT*, 3 April 1931, p. 36.

Chapter 6

1. *NYT*, 1 November 1928, p. 1.
2. *Railway Engineering and Maintenance*, September 1931, p. 803.
3. Griffith, "Single-Phase," pp. 92–93.
4. "Pennsylvania Electrification Links Philadelphia and New York City," *RA*, 94 (25 February 1933):302.
5. *Mutual Magazine*, December 1925, pp. 20, 22; "Electrification Links," p. 283; *RA*, 18 April 1931, p. 297.
6. "Electrification Links," p. 297. The PRR had been experimenting with cab signaling since as early as 1922, mainly in steam territory. By 1931, its New York-Washington and New York-Columbus, Ohio, main lines were equipped for cab signals, as were the locomotives that normally operated over these lines. "Pennsylvania Extends Cab Signaling on Large Mileage," *RA* 89 (9 August 1930):277–79.
7. The canal was the Delaware and Raritan. "Electrification Links." pp. 281–83.
8. Ibid., pp. 279–81; *Railway Engineering and Maintenance*, May 1933, pp. 235–36; J. V. B. Duer, "Construction Procedures for Electrification of Railroads," *RA* 88 (22 March 1930):674–77.
9. *RA*, 3 November 1928, p. 870.
10. *The Pennsylvania Railroad Electrification* (East Pittsburgh: Westinghouse Electric and Manufacturing, n.d.), pp. 15–17; *RA*, 20 June 1931, pp. 1210–11. See also MacLeod, "Recommendations," p. 15.
11. *RA*, 15 July 1947, p. 72; *Who's Who in Railroading* (New York: Simmons-Boardman, 1930), p. 144.
12. J. V. B. Duer, "Electric Locomotives for the Pennsylvania," *Electrical Engineering* 51 (July 1932):489.
13. Ibid., pp. 489–92.
14. Ibid.; J. V. B. Duer, "Pennsylvania Develops Three Types of Electric Locomotives," *RA* 92 (21 May 1932):870–73.
15. *RA*, 16 August 1930, p. 338; 27 June 1931, p. 1269; *PP* I, p. 264. See also *NYT*, 6 September 1931, sec. 2, p. 13.
16. *NYT*, 9 August 1931, p. 7; 1 June 1932, p. 40.
17. Duer, "Three Types of Locomotives," p. 872. See also *NYT*, 27 September 1931, sec. 2, p. 13.
18. *NYT*, 8 April 1934, sec. 2, p. 9; Duer, "Electric Locomotives," p. 488; *PP* I, p. 260. The B1's were equipped with Westinghouse and Allis-Chalmers electrical gear.
19. *Philadelphia Inquirer*, 8 December 1929, p. 13.
20. *Eighty-fourth Annual Report*, 1931, pp. 2, 8–9; *CH*, pp. 584, 628.
21. *NYT*, 30 November 1930, sec. 2, p. 24.
22. Ibid., 19 February 1931, p. 1.
23. *RA*, 21 February 1931, pp. 422–23.
24. Ibid., p. 423.
25. *NYT*, 18 February 1931, p. 1.
26. These bonds were to mature in 1931. Ibid., 13 March 1931, p. 35; *RA*, 6 June 1931, p. 1106.
27. *RA*, 4 April 1931, p. 689; 6 June 1931, p. 1105.
28. *Eighty-fifth Annual Report*, 1932, pp. 2, 9; *Eighty-sixth Annual Report*, 1933, p. 6; *NYT*, 5 February 1932, p. 2.
29. *RA*, 19 March 1932, p. 508; *NYT*, 8 April 1932, p. 37.

30. Loan to PRR: Reconstruction Finance Corporation, Office of the Secretary, "Minutes of the Meetings of the Board of Directors," 1–15 May 1932, 4, pt. 1:98, Record Group 243, 3, National Archives, Washington, D. C. See also 16–31 May 1932, p. 664.

31. *RA*, 19 March 1932, p. 508.

32. "Electrification Links," pp. 271–72; *CH*, pp. 608–10.

33. Duer, "Power Supply," pp. 111–12.

34. Ibid.; *Pennsylvania Railroad Electrification*, pp. 22–24. See also "Electrification Links," pp. 291–93.

35. The 44,000-volt installation at Philadelphia was retained until 1937, at which time the railroad increased the level to 132,000 volts in conjunction with the extension of catenary to Harrisburg. H. C. Griffith, "Extension of the Pennsylvania Railroad Electrified System," *Electrical Engineering* 57 (January 1938):10–11.

36. Duer, "Power Supply," p. 111.

37. Federal Power Commission, *Electric Power in Transportation*, p. 48. See also Duer, "Power Supply," pp. 111–15.

38. *RA*, 17 December 1932, p. 923; "Electrification Links," pp. 291–94. The installation of alternating current at Pennsylvania Station doomed the L5's that had been working there for the past few years. The railroad judged their conversion from d.c. to a.c. to be uneconomical. All were put into storage and then scrapped during World War II. Most of the DD1 locomotives, on the other hand, were transferred to the Long Island Rail Road, which still operated a d.c. system. There they gave many more years of valuable service. The single a.c.-equipped L5 reportedly continued to work in the Philadelphia and Baltimore areas for at least several more years. Westing, *Penn Station*, pp. 170–71, 180–81; Ian S. Fischer to Michael Bezilla, 26 December 1978; 21 February 1979, Piscataway, New Jersey; *RA*, 7 January 1933, pp. 25–26.

39. *NYT*, 17 January 1933, p. 21.

40. Ibid.; *New York Herald-Tribune*, 17 January 1933, p. 3.

41. *Philadelphia Inquirer*, 17 January 1933, p. 2.

42. John H. Dreisbach, "Pennsylvania Engines Discussed by a Practical Man," *Brotherhood of Locomotive Firemen and Enginemen's Magazine* 104 (March 1938):180–81; Austin C. Lescarboura, "The Miracle of the Electric Locomotive," *Travel* 69 (July 1937):4–10.

43. *Brotherhood of Locomotive Firemen* 50 (February 1911):157.

44. *RA*, 21 January 1933, p. 85; *NYT*, 11 February 1933, sec. 2, p. 3; G. H. Burk, "Farewell to Steam," *Railroad Stories* 11 (July 1933):60–65.

45. Temple, "Meadows Division," pp. 75–78; *RA*, 17 September 1915, p. 513.

46. *RA*, 19 January 1929, p. 223; 23 January 1937, p. 203; *Philadelphia Inquirer*, 20 June 1937, p. A2. Although the railroad razed nearly all the structures at Manhattan Transfer and removed some of the track immediately after the abandonment, a wide area in the fill across the Meadows and one of the original interlocking towers (Hudson, formerly Cabin S) still mark the location for the observant rail traveler.

47. Memorandum from G. A. Moriarity, mechanical superintendent, New Haven Railroad, 15 August 1929, CR Thirtieth St.

48. Memorandum from PRR OEE, 1 July 1930, CR Thirtieth St.

49. Duer, "Three Types of Locomotives," p. 871; Bert Pennypacker, "Before the GG1," *Trains* 35 (September 1975):46–48.

50. J. V. B. Duer, "Pennsylvania Railroad Electric Locomotives," Exhibit No. 4, in *Proceedings of the Session of the Association of American Railroads Operations and Maintenance Department, Division V –Mechanical* (Washington: AAR, 1935), p. 62.

51. *NYT*, 13 December 1933, p. 37; Pennypacker, "Before the GG1," p. 49.

52. "Track Tests of Electric Locomotives," *RA* 101 (12 September 1936):374–80; (19 September 1936):412–18.

53. MacLeod, "Recommendations," pp. 6–11; Pennypacker, "Before the GG1," p. 48.

54. *Electrical World*, 4 November 1933, p. 581.

55. "Track Tests," pp. 378–80.

56. *NYT*, 13 December 1933, p. 37; *Electrical World*, 16 December 1933, p. 773; Duer, "Pennsylvania Railroad Electric Locomotives," p. 62.

Chapter 7

1. Duer, "Pennsylvania Railroad Electric Locomotives," p. 62; "Electric Passenger Locomotives," *RA* 100 (15 February 1936):279.

2. Interstate Commerce Commission, Bureau of Safety, *Summary of Accident Investigation Reports*, No. 59, January–March 1934 (Washington: Government Printing Office, 1934), pp. 4–7; Frederick Westing, "Why the Pennsy Modified the P5a Class Electrics," *Railroad History* 130 (Spring 1974): 61–67.

3. General Electric based the EP-3's wheel arrangement on the 20 2–C+C–2 locomotives that it built for the d.c.-equipped Cleveland Union Terminal (New York Central) Railroad in 1929–30. "Track Tests," pp. 379–80.

4. MacLeod, "Recommendations," pp. 7–8; Frederick Westing, *The Locomotives That Baldwin Built* (Seattle: Superior Publishing, 1966), p. 159.

5. J. W. Horine and H. S. Ogden, "The Pennsylvania Railroad Class GG-1 Electric Locomotives," *TAIEE* 79–2 (May 1960):107; Westing, *Locomotives That Baldwin Built*, p. 159.

6. Horine and Ogden, "Class GG-1," pp. 107–10; "Electric Passenger Locomotives," pp. 279–82.

7. Warner, "Motive Power Development" (October 1924):29.

8. "Track Tests," pp. 378–80.

9. W. D. Bearce, "The Pennsylvania Railroad Electrification: New York-Washington," *General Electric Review* 39 (March 1936):144.

10. *RA*, 17 November 1934, p. 625; *NYT*, 11 November 1934, sec. 2, p. 11.

11. Raymond Loewy, *Never Leave Well Enough Alone* (New York: Simon and Schuster, 1951), pp. 135–41. The concept of the welded carbody was soon widely adopted for both diesel and electric locomotive construction, as well as for other kinds of rolling stock.

12. Many PRR passenger diesels appeared in red paint (closely resembling the tuscan red of the road's passenger equipment) through the 1950s. The GG1's themselves have worn numerous paint schemes through their long lives. Richard H. Pfeiffer, "Forty Years Young," *PC Railroader* 2 (March–April 1974):21–35.

13. F. W. Hankins to R. G. B., 16 June 1936, CR Thirtieth St.

14. "Electric Passenger Locomotives," p. 278; Pennypacker, "Before the GG1," p. 50.

15. *RA*, 15 July 1933, p. 135.

16. RFC "Minutes," 18, pt. 1; 22–30 June 1933, 17, pt. 3; 1–13 July, 1933, 18, pt. 1; 22–29 July 1933, 18, pt. 3; *NYT* 29 July 1933, p. 17. By 1934, salaries of the PRR's chief officers averaged only 56% of their 1929 levels. *NYT*, 10 April 1935, p. 31.

17. *Eighty-seventh Annual Report*, 1934, p. 6.

18. Harold L. Ickes, *Back to Work: The Story of the PWA* (New York: Macmillan, 1935), p. 152; *The Secret Diary of Harold L. Ickes*, 3 vols. (New York: Simon and Schuster, 1953),1:81.

19. *Secret Diary*, p. 115.

20. *RA*, 11 November 1933, p. 695; 6 January 1934, pp. 15–16.

21. Ibid., 6 January 1934, pp. 15–16; *NYT*, 30 December 1933, p. 19; 31 January 1934, p. 12.

22. *NYT*, 31 January 1934, p. 12. See also Herman B. Buyer, "Labor Requirements of a Railroad Electrification Program," *Monthly Labor Review* 43 (September 1936):586–90.

23. The PWA eventually allowed the PRR to use part of its loan to pay this $700,000, although GE and Westinghouse had done the work before the loan was granted. C. J. Maxcy, chief accountant of the PWA, to F. C. Wright, Transportation Loan division of the PWA, 13 March 1935, Public Works Administration, Records of the Accounting division, Record Group 135.108, National Archives, Washington, D.C.

24. Griffith, "Single-Phase," pp. 92, 93–95; *Electrical World*, 14 July 1928, p. 85; 27 October 1928, p. 860; 29 December 1928, p. 1317.

25. *CH*, pp. 618–19; *Railway Engineering and Maintenance*, February 1936, p. 74. To this day, the Baltimore and Potomac tunnel has been neither replaced nor enlarged.

With its double track, restricted clearances, and steep (1.4 %) westbound grade, it remains one of the most troublesome bottlenecks on Amtrak's New York-Washington main line.

26. *RA*, 10 March 1934, p. 352; Griffith, "Single-Phase," p. 94.

27. "Railroad Electrification As Seen from the Baltimore Angle: An Interview with Herbert A. Wagner," *Electrical World* 92 (29 December 1928):1303. Wagner was president of Consolidated Gas and Electric.

28. The Potomac Electric Power Company owned the equipment at Benning, but Consolidated Gas supplied the current. Safe Harbor's point of delivery to the PRR was at Perryville, Maryland. *RA*, 17 October 1931, p. 610; Griffith, "Single-Phase," p. 94.

29. "Electrification Links," pp. 295–96.

30. *RA*, 27 April 1935, p. 651; *NYT*, 21 September 1935, pp. 1–2. Atterbury never attended another board meeting after his initial hospital stay. *CH*, p. 645. Atterbury: quoted in *RA*, 27 April 1935, p. 651.

31. *RA*, 27 April 1935, p. 651: "Pennsylvania Railroad" (May 1936):68, 72, 77; *CH*, pp. 645-72.

32. The PRR renumbered its first GG1 in order to more easily identify it as the first in a series of identical locomotives of that class. The R1 was given No. 4899, since 647, 649.

33. Quoted in ibid.

34. Quoted in ibid.

35. Quoted in ibid.

36. *RA*, 2 February, 1935; *NYT*, 29 January 1935, p. 23.

37. *RA*, 16 February 1935, p. 278; 13 April 1935, p. 584.

38. Ibid., 2 February 1935, p. 197.

39. Ibid., 25 May 1935, p. 833; 15 June 1935, p. 944; *NYT*, 21 May 1935, p. 32.

40. *RA*, 21 April 1934, p. 589.

41. Ibid., 27 April 1935, p. 663; 14 September 1936, p. 345; 25 April 1936, p. 704; 3 October 1936, p. 488.

42. Ibid., 1 July 1933, p. 73; 26 August 1933, p. 322; 5 May 1934, p. 673. Stokers: Initially the railroad installed stokers on locomotives having a firing rate greater than 4,000 pounds per hour. F. W. Hankins to Martin W. Clement, 14 July 1931, Package No. R5650, CR Merion Ave; Westing, "K4s," pp. 50–51.

43. "Evolution from Steam to Electric Traction," *RA* 107 (23 September 1939):446, 449.

44. Charles H. Mathews, Jr., general manager of passenger traffic, to Walter S. Franklin, vice-president in charge of traffic, 17 August 1936, CR Thirtieth St.; J. F. Deasy to Martin W. Clement, 9 October 1936, CR Thirtieth St.

45. *CH*, pp. 647–649.

46. Broad Street Station was finally demolished in 1952 to make way for the PRR-sponsored Penn Center office complex. By that time, too, the Reading no longer posed much of a threat to the Pennsylvania's dwindling New York-Philadelphia passenger traffic. Ibid., p. 611; *RA*, 5 May 1952, pp. 92–94.

47. *Ninetieth Annual Report*, 1937, p. 8; *RA*, 30 January 1937, p. 234; *Philadelphia Inquirer*, 28 January 1937, p. 1.

48. *RA*, 30 January 1937, p. 234; "Electrification," pp. 15–16.

49. Griffith, "Extension," pp. 11–14; *NYT*, 23 September 1939, p. 28.

50. *Philadelphia Inquirer*, 16 January 1938, p. A-2.

51. *Harrisburg Patriot*, 17 January 1938, p. 1.

52. *RA*, 23 April 1938, p. 746.

53. "Memorandum of the Meeting of Committee Set Up by F. W. Hankins on the Subject of Results Obtained by Electrified Service, 31 October 1938," CR Thirtieth St.

54. Electrification saved the railroad $886,124 in freight operations, $4,985,043 in passenger operations, and $1,609,725 in multiple-unit operations. "Eastern Electrification Traffic for the Year 1938," n.d., CR Thirtieth St. See also *NYT*, 10 April 1935, p. 31.

55. The PRR outshopped its first M1 in 1923 as an experimental unit. Quantity production began in 1926. "Minutes of the Meeting on Results Obtained from Elec-

trified Service, 21 December 1938," CR Thirtieth St. Maintenance and repair costs of all PRR a.c. locomotives averaged 11.3 cents per mile. "Gross Costs of Repairs to A.C. Type Electric Locomotives," n.d., CR Thirtieth St.

56. Electric locomotives underwent monthly inspections either at Enola or Wilmington. Class repairs were performed at Wilmington only. PRR OEE Memorandum, 31 December 1940, CR Thirtieth St. See also Duer, "What Electric Operation Is Doing," p. 687.

57. PRR OEE Memorandum, 31 December 1940, CR Thirtieth St.; Duer, "What Electric Operation Is Doing," p. 687.

58. Griffith, "Electric Locomotive Operation," pp. 233–34.

59. Ibid.; PRR OEE Memorandum, 16 February 1943, CR Thirtieth St.

60. H. D. A. to H. C. Griffith, PRR electrical engineer, 24 March 1939, CR Thirtieth St.; *CH* p. 659.

61. Griffith, "Electric Locomotive Operation," p. 234.

62. Quoted in Lescarboura, "Miracle of the Electric Locomotive," p. 44.

63. "Trouble Shooter for a Railroad," *Popular Mechanics* 37 (January 1937):58–61; MacLeod, "Recommendations," p. 2; *RA*, 25 February 1933, p. 293. For examples of specific incidents involving animals, see "High Tension Story," *Pennsy* 1 (August 1952):8–10; *NYT*, 11 January 1940, p. 2; and *Philadelphia Inquirer*, 21 July 1947, p. 3.

64. "Blizzard of '58," *Pennsy* 7 (March-April 1958):8–10; Horine and Ogden, "Class GG-1," p. 8. The snowflakes adversely affected only the GG1's. Upon inquiry at the local weather bureau, the railroad learned that this type of "diamond" snow was extremely rare and had not fallen in the Northeast in the last 25 years. The flakes were so fine that they could penetrate crevices through which water could not even pass and were particularly numerous near ground level. Thus the GG1's, with their low air inlets, were vulnerable, while the P5a's and diesels, with inlets located higher on the car bodies, were not.

65. Bearce, "New York-Washington" (February 1936):100.

66. *RA*, 12 January 1929, p. 168; 23 March 1929, p. 680; *SR*, pp. 301–10.

67. MacLeod, "Recommendations," p. 38; Martin W. Clement, "Railroads in the War," *JFI*, 242 (November 1946):345–54; A. C. Kalmbach, "Epoch of Electrification," *Trains* 6 (April 1946):40; *CH*, pp. 681–88.

68. Duer, "What Electric Operation Is Doing," pp. 685–86.

69. Kalmbach, "Epoch," p. 40; "The Pennsy's Predicament," *Fortune* 37 (March 1948):202.

Chapter 8

1. "Executive's View," p. 1137.

2. *Pittsburgh Press*, 4 June 1930, p. 24; *NYT*, 5 June 1930, p. 13.

3. Federal Power Commission, *National Power Survey*, p. 20; D. R. MacLeod, "PRR Pittsburgh Electrification Study," n.d., p. 3, GE Erie.

4. The average freight train operating on the main lines of the Middle and Pittsburgh divisions in 1930 weighed 4,000 gross tons, compared to an average freight train weight of 2,950 gross tons in electrified territory. D. R. MacLeod, "PRR Pittsburgh Electrification Study," p. 3.

5. Federal Power Commission, *National Power Survey*, pp. 3–24.

6. Gibbs and Hill and the Pennsylvania Railroad, "Study of Proposed Electrification, Harrisburg to Pittsburgh," pp. 2–3, CR Thirtieth St. This study contained a summary of the recommendations of the 1936–38 Gibbs and Hill report.

7. George Gibbs to F. W. Hankins, 17 June 1938, CR Thirtieth St.

8. PRR OEE Memorandum, 24 May 1937, CR Thirtieth St.; J. Stair, Jr., PRR electrical engineer, to H. T. Cover, PRR chief of motive power, 26 March 1946, CR Thirtieth St.

9. MacLeod, "Pittsburgh Study," pp. 1–6.

10. Gibbs and Hill estimated that electrification would save the PRR $1,500,000

annually in helper service costs, $1,200,000 in freight operations, and $950,000 in passenger operations, based on 1939 traffic levels. "Proposed Electrification, Harrisburg to Pittsburgh," pp. 2–3, 7.

11. *Pittsburgh Post-Gazette,* January 1944, p. 9; *NYT,* 1 January 1944, p. 20. See also G. M. Davis of General Electric, and W. A. Brecht of Westinghouse, to H. W. Jones, PRR chief of motive power, 28 August 1944, CR Thirtieth St.; E. L. B. to Martin W. Clement, 30 August 1945, p. 201; Gibbs and Hill and the Pennsylvania Railroad, "Supplement to Proposed Electrification, Harrisburg-Pittsburgh," 1945, p. 201, CR Thirtieth St.

12. *One Hundredth Annual Report,* 1947, p. 1; *Ninety-fourth Annual Report,* 1941, p. 2.

13. *One Hundredth Annual Report,* pp. 1–4.

14. *Ninety-seventh Annual Report,* p. 3. See also *One Hundred Second Annual Report,* 1949, p. 5.

15. *One Hundredth Annual Report,* p. 2; *Ninety-fourth Annual Report,* p. 2; "The Pennsy's Predicament," (March 1948):85–93.

16. "The Pennsy's Predicament" (March 1948):201.

17. *One Hundred First Annual Report,* 1948, p. 1; *One Hundred Second Annual Report,* pp. 1–2.

18. *PP* I, pp. 216–44. See also "The Pennsy's Predicament" (March 1948):202; *One Hundred Seventh Annual Report,* 1954, p. 116.

19. Charles Kerr, Jr., "What Diesels Mean to Railroads," *RA* 134 (6 April 1953):69–71.

20. *One Hundred First Annual Report,* pp. 13–15.

21. *One Hundred Second Annual Report,* p. 6.

22. *One Hundred Third Annual Report,* 1950, p. 2.

23. U. S. Congress, Senate, Committee on Interstate and Foreign Commerce, *Hearings Before the Subcommittee on Surface Transportation of the Committee on Commerce,* Statement of James M. Symes, 85th Cong., 2d sess., 1958, p. 77.

24. The PRR began regearing the P5a's from 90 miles per hour to 70 miles per hour in 1934. Box cabs were first to receive regearing, since the railroad still planned to use the modifieds in occasional passenger service. F. W. Hankins to R. G. B., 16 June 1936, CR Thirtieth St.; J. F. Deasy to F. W. Hankins, 17 June 1938, CR Thirtieth St.

25. J. M. Symes to M. W. Clement, 3 December 1948, CR Thirtieth St. The PRR motorized the nonpowered axles of P5a No. 4702 (thus producing the only P5b), but the improved adhesion resulting from the alteration was not considered worth the extra cost. MacLeod, "Recommendations," pp. 10–11.

26. "Proposed 7500 hp Electric Locomotives," PRR OEE, 1946, pp. 1–10, CR Thirtieth St.

27. "Outline Specifications for a 7500 hp Single-phase Freight Locomotive," PRR OEE, 1947, CR Thirtieth St.

28. J. M. Symes to M. W. Clement, 3 December 1948, CR Thirtieth St.

29. M. W. Clement to Charles E. Wilson, 8 June 1948, CR Thirtieth St.

30. L. W. Ballard to H. T. Cover, 19 November 1948, GE Erie; F. D. Brown of Westinghouse, to E. J. Lamneck, PRR general purchasing agent, 12 October 1948, CR Thirtieth St.

31. Griffith, "Electric Locomotive Operation," pp. 230–32.

32. PRR OEE Memorandum, 19 August 1949, CR Thirtieth St. This document recounts details of two meetings at which PRR, Westinghouse, and GE representatives discussed specifications of the proposed locomotives, especially regarding the railroad's weight, clearance, and safety requirements. The PRR had placed its order for the locomotives earlier in the year, however. See *RA,* 26 March 1949, p. 670.

33. F. D. Gowans, B. A. Widell, and A. Bredenberg, "A New Electric Locomotive for the Pennsylvania Railroad," *TAIEE* 71–2 (January 1952):27–36; C. C. Whittaker and W. M. Hutchison, "The Pennsylvania Railroad Ignitron Rectifier Locomotive," *TAIEE* 71–2 (January 1952):37–42; A. C. Monteith, "The Ignitron Locomotive and Its Effect on Electrification," *Westinghouse Engineer* 10 (July 1950):187–88.

34. Gowans, Widell, and Bredenberg, "New Electric Locomotive," pp. 28–36; MacLeod, "Recommendations," p. 29. General Electric built an additional two-unit set to

demonstrate on the New Haven and the Great Northern railroads. The PRR acquired the set early in 1952, after it had completed its demonstrations, and gave it Nos. 4943–4944. *General Electric Review*, July 1952, p. 40; September 1952, pp. 43–46.

35. MacLeod, "Recommendations," pp. 2–3. See also A. C. Monteith, "Rectifier-Type Locomotives Now Being Built for the Pennsylvania," *RA* 128 (March 1950):589. The PRR had placed a 60,000-pound axle-loading limit on the new electrics. Gowans, Widell, and Bredenberg, "New Electric Locomotive," p. 27.

36. Gowans, Widell, and Bredenberg, "New Electric Locomotive," pp. 30–32.

37. Ibid., p. 27.

38. Murray, "Mainline Electrification," p. 519; Monteith, "Rectifier-Type," p. 589; *Railway and Engineering Review*, 28 March 1908, p. 266; 11 April 1908, p. 304.

39. Whittaker and Hutchison, "Ignitron Rectifier Locomotive," p. 37; Richard W. Dodge, "The Rectifier Railway Car – AC Supplied, DC Driven," *Westinghouse Engineer* 9 (November 1949): 180.

40. L. J. Hibbard, C. C. Whittaker, and E. W. Ames, "Rectifier-Type Motive Power for Railroad Electrification," *TAIEE* 69-1 (1950):522; Dodge, "Railway Car," p. 179.

41. Hibbard, Whittaker, and Ames, "Rectifier-Type Motive Power," pp. 519–22.

42. Dodge, "Railway Car," p. 180; Harry F. Dimmler, "Pennsy's New Juice Jacks," *Trains*, 14 (February 1954):19. See also Hibbard, Whittaker, and Ames, "Rectifier-Type Motive Power," p. 591.

43. W. E. Kelley, "Historical Summary, Performance, and Future of Penn Central Company Railroad Electrification," in *Conference on Performance of Electrified Railways* (London: Institution of Electrical Engineers, 1968), p. 69; Whittaker and Hutchison, "Ignitron Rectifier Locomotive," p. 37.

44. Monteith, "Rectifier-Type," p. 591.

45. Ibid.; Monteith, "Effect," pp. 187–88; Whittaker and Hutchison, "400,000 Gross Ton-Miles Per Train-Hour," *RA* 132 (30 June 1952):69.

46. *RA*, 7 January 1952, p. 15; Whittaker and Hutchison, "Ignitron Rectifier Locomotive," pp. 38–41. The cost of these experimental locomotives would have been considerably less had they been mass produced. Neither builder quoted a figure for quantity production, however. And because of inflation's impact during the late 1940s, a comparison between the cost of these experimentals and the cost of, say, a GG1, would be misleading. "An Approximation of the Losses Sustained by the PRR Due to the Purchase of E2b, E2c, and E3b Experimental Freight Locomotive Units," PRR OEE, 1962(?), CR Thirtieth St.

47. S. V. Smith, "The Pennsylvania's New Electric Freight Locomotives: A Comparison," *RA* 132 (10 March 1952): 69. J. V. B. Duer, formerly chief electrical engineer, was appointed assistant to the vice-president of operations of the PRR in 1940 and retired from that post in 1947. He received the Franklin Institute's George R. Henderson medal in 1949 in recognition of his contribution to railway electrification. Because of subsequent reorganizations of the engineering departments and the low priority the railroad accorded to electrification after the war, neither S. V. Smith nor any of his successors exercised the same degree of influence and authority as Duer. *Electrical Engineering*, October 1949, p. 903; *RA*, 5 July 1947, p. 72.

48. Whittaker and Hutchison, "400,000," pp. 68–69.

49. F. D. Brown, "Some Application Phases of the Ignitron Rectifier Locomotives of the Pennsylvania Railroad," *TAIEE* 73-2 (July 1954):128–34.

50. PRR OEE Memorandum, 18 April 1952, CR Thirtieth St.

51. S. V. Smith, "Comparison," p. 68; Dimmler, "Juice Jacks," pp. 15–22; Henry B. Comstock, "DC Battles AC to Pull the Heavy Loads," *Popular Science* 160 (May 1952):96–100, 270.

52. H. T. Cover to J. Stair, Jr., 24 September 1954, CR Thirtieth St. W. E. Lehr, who succeeded Cover as PRR chief of motive power, also recounted the failures of these units in a letter to D. E. S., 29 July 1964, CR Thirtieth St.

53. S. V. Smith to W. E. Lehr, 24 July 1964, CR Thirtieth St. Fires arising from the ignitron's complicated cooling system also plagued the locomotives. The Pennsylvania's engineers had foreseen a fire danger at the time of the units' construction and insisted that Westinghouse provide an extinguishing mechanism whereby the equipment spaces could be flooded with carbon dioxide foam. The foam did nothing to

prevent fires from starting, of course. H. T. Cover to L. A. Lester, Westinghouse, 1 April 1949, CR Thirtieth St.

54. Ben F. Anthony, General Electric applications engineer, to Michael Bezilla, 28 November 1977.

55. S. V. Smith to W. E. Lehr, 24 July 1964, CR Thirtieth St.

56. L. W. Ballard to D. S. Onnen, Locomotive and Power Equipment department superintendent, General Electric, 10 November 1952, GE Erie.

57. L. E. Gingerich, PRR chief mechanical officer, to Allen J. Greenough, PRR vice-president of Transportation and Maintenance, 3 March 1959, CR Thirtieth St.

58. H. T. Cover to F. C. Ruling, Atlantic division manager of General Electric, 6 January 1954, GE Erie.

59. "Study of PRR Co. Electric Locomotive Maintenance by GE Co.," 1954, GE Erie.

60. "An Approximation of the Losses"; Kelley, "Summary," p. 68.

61. *Westinghouse Engineer,* January 1953, p. 38; *RA,* 15 March 1954, pp. 66–68; *General Electric Review,* September 1955, pp. 57, 59–60; May 1958, pp. 34–36.

62. *RA,* 7 June 1954, p. 16.

63. L. W. Ballard to T. F. Perkinson, 30 September 1949, GE Erie; D. R. Mac-Leod: quoted in discussion appended to Whittaker and Hutchison, "Ignitron-Rectifier Locomotive," p. 43.

64. "Minutes of Rectifier Committee Meeting No. 1, 26–27 January 1953," GE Erie.

65. *One Hundred Eighth Annual Report,* 1955, p. 2; Herrymon Maurer, "New Signals for the Pennsy," *Fortune* 52 (November 1955):159–60. See also *One Hundred Eleventh Annual Report,* 1958, p. 2.

66. *One Hundred Tenth Annual Report,* 1957, p. 5.

67. Senate, *Hearings,* p. 108.

68. Ibid., pp. 77–79.

69. *Philadelphia Inquirer,* 9 May 1955, p. 30; *Traffic World.* 14 May 1955, pp. 39–40; *RA,* 16 May 1955, p. 12.

70. Gibbs and Hill, "An Engineering and Economic Study of Motive Power Replacement with an Evaluation of Electrification," 1955, Preface, CR Thirtieth St.

71. Ibid., pp. 3–6.

72. PRR Research and Development Department, "Equipment Study–Electric Locomotives," 1957, pp. 1–2, CR Thirtieth St.; General Electric, "Pennsylvania Railroad Future Motive Power in Electrified Territory," 1959, pp. 1–5, GE Erie.

73. James P. Newell to Allen J. Greenough, 20 September 1957, CR Thirtieth St.

74. "Summary of Circumstances Relative to Purchase of FF2's," PRR OEE, 1957, CR Thirtieth St.; Allen J. Greenough to H. T. Cover, 3 April 1957, CR Thirtieth St.; Frederick Westing, "What's New Under Pennsy Pantographs," *Trains* 18 (June 1958):45–49.

75. PRR OEE Memorandum, 30 October 1957, CR Thirtieth St.; Memorandum from J. J. Clutz, PRR director of research, 17 June 1957, GE Erie.

76. James P. Newell to R. S. Dickson, president of R. S. Dickson and Company, 8 April 1960, CR Thirtieth St. Newell summarized the builders' reports in this letter. R. S. Dickson and Company was a coal marketing firm.

77. General Electric, "Future Motive Power," pp. 1–3, 15.

78. Allen J. Greenough to Joseph T. Berta, Pittston Clinchfield Coal Sales Corporation, 7 September 1961, CR Thirtieth St.

79. Allen J. Greenough to James P. Newell, 8 May 1960, CR Thirtieth St.

80. *One Hundred Thirteenth Annual Report,* 1960, p. 10; *RA,* 7 November 1960, p. 18.

81. Quoted in *RA,* 17 November 1960, p. 18. Greenough served as president until his company merged with the New York Central in 1968. A managerial reorganization in the 1950s had made the PRR's chairman of the board the company's most powerful executive, however. James Symes served as chairman until 1963, when he was succeeded by a former Norfolk and Western official, Stuart T. Saunders. Saunders later became the Penn Central's first board chairman.

82. J. C. Brown and J. W. Horine, "Application of Silicon Rectifiers on Electric Locomotives in the United States," *IEEE Transactions on Applications and Industry* 82 (May

1963):133–35; "Pennsy Electrics Point the Way," *RA* 158 (12 April 1965):18–19.

83. Kelley, "Summary," p. 67; Pennypacker, "Before the GG1," pp. 50–51. One P5a has been preserved. In 1965, the PRR donated No. 4700, one of the first pair of prototype P5's, to the National Museum of Transport in St. Louis, Missouri. It is only the third PRR electric locomotive that is still preserved. A DD1 (Nos. 3936–37) and a B1 (No. 4756) are at the Railroad Museum of Pennsylvania at Strasburg, Pa.

84. Kelley, "Summary," p. 69; S. V. Smith, "Modern Efficient Silicon Rectifier-Type Multiple-Unit Cars for Philadelphia Area Commuter Service," *IEEE Transactions on Applications and Industry* 83 (November 1964):344. See also *RA*, 24 April 1961, p. 42; "Pennsy Electrics," pp. 18–19.

85. Kelley, "Summary," pp. 70–72; Bert Pennypacker, "PRR vs. Plane, Car, Bus," *Trains* 28 (November 1967):18–28.

86. Kelley, "Summary," pp. 70–71; *RA*, 20 January 1969, p. 8; Pennypacker, "PRR vs. Plane, Car, Bus," pp. 18–28.

Epilogue

1. Edward T. Myers, "Ready or Not," *Modern Railroads* 31 (August 1976):50–53; Luther H. Miller, "Northeast Corridor: Operation Fix-Up Gets Under Way," *RA* 178 (9 May 1977):35–36, 43.

2. Myers, "Ready or Not," pp. 50–53; Michael Bezilla, "The Electrification That Might Have Been–And Might Still Be," *Trains* 38 (March 1978):30–34.

3. An excellent survey of the problems and prospects of railroad electrification in the United States in recent years is *Railroad Electrification: The Issues*, Transportation Research Board Special Report No. 180 (Washington: National Academy of Sciences, 1977). It addresses the specific challenges facing Amtrak and Conrail, as well as more general questions. See also William D. Middleton, "Railroad Electrification and Energy Conservation," *Traffic Quarterly* (July 1978):383–93, and Robert Roberts, "Status Report: Electrification," *Modern Railroads* 33 (June 1978):56–59.

A Note on the Sources

The chief source of Pennsylvania Railroad documents used in this study was a collection of engineering records located in the Consolidated Rail Corporation's Office of Engineering and Research, Thirtieth Street Station, Philadelphia. A vast collection of PRR documents is housed in Conrail's Merion Avenue Records Center in West Philadelphia. Unfortunately, the Center is little more than a large warehouse, and the index and retrieval systems for the materials there are extremely inadequate. Consequently, only limited use could be made of the huge quantity of documents stored there. Another collection of PRR electrification records (the exact nature of which is unknown) was recently acquired by the Commonwealth of Pennsylvania for eventual deposit at the Railroad Museum of Pennsylvania at Strasburg. As of this writing, these records have not yet been catalogued and are not open for research.

General Electric, like Conrail, has on file engineering records dating back about 40 years. Most of the electrification files of GE's rival, Westinghouse, were destroyed a number of years ago in a flood. The remaining files were discarded when Westinghouse ceased building heavy electric locomotives. The records of the consulting firm of Gibbs and Hill, still in business 70 years after its founding, were not available for research.

Few government records pertaining to PRR electrification exist. Documents relating to the railroad's Reconstruction Finance Corporation loan, reported to be on file at the National Archives in Washington, are missing and presumed lost. Most of the files of the Public Works Administration's Transportation Loan division were accidentally destroyed during the 1940s. A few relevant documents can be found in the files of the PWA's Accounting division, however.

Fortunately, a plentiful number of articles (in most cases written by individuals directly involved in electrification) are to be found in the various trade and professional engineering journals. These partially make up for the dearth of corporate records for the years prior to World War II. Newspapers and trade publications also carried brief news items from time to time concerning electric traction activities on the Pennsylvania. Only those periodicals giving consistent coverage over relatively long periods of time are cited in the Notes.

Secondary sources on the PRR electrification are fairly numerous, but most are of limited value to scholars. Rail enthusiasts have authored the majority of these works, since professional historians have done practically no research and writing on the subject of railroad electrification. In fact, the only historical study on railroad electrification that is worth mentioning is Carl Condit's slim volume, *The Pioneer Stage of Railroad Electrification* (1977), which treats only the very early years of electric traction. This has now been published as part of a larger work by the same author, *Port of New York* (Chicago: University of Chicago Press, 1980). By far the most comprehensive study is William D. Middleton's *When the Steam Railroads Electrified* (1974). While Middleton's is a popular rather than a scholarly work, it is nonetheless an excellent general survey of the technological evolution of American railroad electrification. However, its broad scope necessarily prevents it from giving detailed consideration to developments on the Pennsylvania.

One of the weaknesses of many secondary works is the lack of documentation by which their content may be judged for accuracy. Another is their preoccupation—not unexpected, in view of their intended readership—with a pictorial approach to electrification. And even many of the more factual, nonpictorial sources further restrict their usefulness by confining themselves almost exclusively to fairly technical discussions of motive power. Still, a few secondary books and articles do provide valuable information. Most of these are cited in the Notes.

Scholars have been scarcely more productive in regard to the general history of the Pennsylvania Railroad. The standard documentation of the road's past is Burgess and Kennedy's outdated (1949) *Centennial History of the Pennsylvania Railroad Company.* The railroad itself commissioned this work, yet the authors' objectivity and candor stand in sharp contrast to the public relations veneer that characterizes many other corporate histories. *Centennial History* pays only fleeting attention to technological developments, and electrification is therefore dealt with all too briefly. H. W. Schotter's *Growth and Development of the Pennsylvania Railroad Company* is an even older (1927) work and contains little that cannot be found in *Centennial History.* The only modern history of note is Robert Sobel's *The Fallen Colossus* (1977), an overview of the histories of the Pennsylvania and the New York Central railroads from their nineteenth century origins through the Penn Central debacle. Unfortunately, this study is filled with interpretive inaccuracies and misstatements of fact, especially in the realm of technology.

Appendix A
Locomotive Classification
by Wheel Arrangement

Railroads and their equipment suppliers since the nineteenth century used the Whyte system to describe general categories of steam locomotives. This system used a numerical designation based upon the number and position of a locomotive's wheels. A locomotive in widespread use in the nineteenth cenetury, for example, had four guiding wheels, four driving wheels, and no trailing wheels. The Whyte system thus identified it as a 4–4–0. When some railroads enlarged this engine and placed a two-wheel trailing truck behind the drivers to support a bigger firebox, a new classification came into being, the 4–4–2. As locomotives grew in size and power, many different kinds of wheel arrangements appeared: 2–8–0, 2–8–2, 4–6–2, and 4–8–2, to name only some of the more popular types.

When electric locomotives made their debut on American railways, they, too, were at first classified according to the Whyte system. However, this scheme failed to distinguish between driving (powered) and idler (nonpowered) wheels. The drivers were easily identifiable in a steam locomotive listing, but not so with electrics, which had greater flexibility in the location of powered and nonpowered wheels. Therefore, in the early 1900s, American railroads borrowed from the European practice of counting axles rather than wheels and using numbers to denote nonpowered axles and letters to denote powered ones. No provision existed (nor was one necessary) to denote the absence of an axle. Thus PRR electric locomotive No. 10,003, formerly termed a "4–4–0," now became a 2–B. Had it possessed only one driving axle, it would have been a 2–A. Three driving axles would have made it a 2–C. The "–" between letters and /or numbers designated separate trucks. The New Haven's Class EP–1, riding on two separate trucks each having two powered axles, was a B–B. A "+" stood for articulated trucks, that is, trucks hinged together. The Pennsylvania's No. 10,002, riding on two four-wheel articulated trucks, was thus a B+B. The DD1's 2–B+B–2 arrangement meant two nonpowered axles, followed by two articulated trucks having two powered axles each, followed by two more nonpowered axles.

By the late 1920s, this number-and-letter system had been adopted by all American electric locomotive manufacturers and their railroad customers. Each road continued to give a particular wheel arrangement a certain class on the motive power roster, of course. (See Appendix B.) Hence the PRR designated its 2–C+C–2 as Class GG1, while the New Haven gave the same arrangement Class EP–3 or EP–4. When diesel locomotives made their appearance, they were classified in the same manner as electrics. The most common diesel types had B–B or C–C wheel arrangements.

Appendix B
Pennsylvania Railroad Steam and
Electric Locomotive Classification

The Pennsylvania Railroad developed its own system of classifying locomotives by identifying each of the Whyte arrangements by letter. Generally speaking, after a reorganization of this system in the 1890s, the further the alphabetical listing progressed, the larger and more modern the locomotives became. Class A engines were small 0-4-0 switcher types, while Class M steamers were powerful 4-8-2 road locomotives. In many cases, the PRR patterned its electric locomotive class designations after those of its steam fleet. In 1932, when its 1-D-1 electric freighter appeared, the PRR termed it Class L6. Steam Class L1 was a 2-8-2, which is the Whyte version of 1-D-1. Steam Class G was a 4-6-0. The GG1 electric embodied two 4-6-0 arrangements placed back to back, for a 2-C+C-2 designation. Occasionally, no steam class existed to provide a precedent, as in the case of the P5a, a 2-C-2 or 4-6-4.

This policy changed after World War II. Class E3b, for example, denoted an electric locomotive having three sets of B trucks. Class E-44 stood for electric locomotive, 4,400 horsepower. In the case of the E-44, the railroad simply adopted the manufacturer's designation for the unit.

The numbers that followed the alphabetical listings usually identified the various subclasses of locomotives having a common wheel arrangement and signified the existence of major mechanical or design differences among the subclasses. The Pennsylvania classified its first 4-4-2 as E1, for instance, but later subclasses included E2, E3, E5, and E6. The small suffix letter "s" (as in K4s) until the late 1920s denoted steam locomotives having superheaters. Other suffix letters, such as "a" and "b," in most cases identified relatively minor but still significant alterations performed on steam or electric engines within a particular subclass: I1s, I1sa; M1, M1a; P5, P5a, P5b.

Appendix C

Pennsylvania Railroad Electric Locomotives

Class	Wheel Arrangement	Horsepower[1] (continuous)	Quantity	Construction Dates
AA1[2]	B–B	–	2	1905
Odd-D[3]	2–B	750	1	1907
DD1[4]	2–B+B–2	3,160	33	1909–11
FF1	1–C+C–1	4,000	1	1917
L5	1–D–1	3,040	24	1924–28
B1	C	570	28	1926–35
O1	2–B–2	2,000	8	1930–31
P5	2–C–2	3,750	92	1931–35
L6	1–D–1	2,500	3	1932–33
R1	2–D–2	5,000	1	1934
GG1	2–C+C–2	4,620	139	1934–43
DD2	2–B+B–2	5,000	1	1938
E2b	B–B	2,500	6	1951
E2c	C–C	3,000	2	1951
E3b	B+B+B	3,000	2	1951
FF2	1–C+C–1	3,300	7	1926–29
E-44	C–C	4,400	66	1960–63

[1]Because many of the PRR's electric locomotives underwent numerous modifications over the years, and propulsion equipment varied slightly from one subclass to another, horsepower figures are approximate.
[2]Class AA1 consisted of experimental No. 10,001 (1,400 horsepower) and No. 10,002 (1,240 horsepower).
[3]Class Odd-D was experimental No. 10,003.
[4]Class DD1 includes 2 two-unit Class Odd-DD locomotives, which were prototypes of the DD1's.

Index

●●●